Vahid Pirouzfar, Yeganeh Eftekhari, Chia-Hung Su
Pinch Technology

I0050971

Also of Interest

Process Intensification.
Breakthrough in Design, Industrial Innovation Practices,
and Education
Jan Harmsen, Maarten Verkerk, 2020
ISBN 978-3-11-065734-0, e-ISBN (PDF) 978-3-11-065735-7,
e-ISBN (EPUB) 978-3-11-065752-4

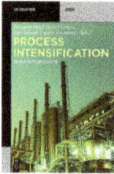

Process Intensification.
Design Methodologies
Fernando Israel Gómez-Castro, Juan Gabriel Segovia-Hernández
(Eds.), 2019
ISBN 978-3-11-059607-6, e-ISBN (PDF) 978-3-11-059612-0,
e-ISBN (EPUB) 978-3-11-059279-5

Product-Driven Process Design.
From Molecule to Enterprise
Edwin Zondervan, Cristhian Almeida-Rivera, Kyle Vincent Camarda,
2020
ISBN 978-3-11-057011-3, e-ISBN (PDF) 978-3-11-057013-7,
e-ISBN (EPUB) 978-3-11-057019-9

Multi-level Mixed-Integer Optimization.
Parametric Programming Approach
Styliani Avraamidou, Efstratios Pistikopoulos, 2022
ISBN 978-3-11-076030-9, e-ISBN (PDF) 978-3-11-076031-6,
e-ISBN (EPUB) 978-3-11-076038-5

Process Systems Engineering.
For a Smooth Energy Transition
Edwin Zondervan (Ed.), 2022
ISBN 978-3-11-070498-3, e-ISBN (PDF) 978-3-11-070520-1,
e-ISBN (EPUB) 978-3-11-070531-7

Integrated Chemical Processes in Liquid Multiphase Systems.
From Chemical Reaction to Process Design and Operation
Matthias Kraume, Sabine Enders, Anja Drews, Reinhard Schomäcker,
Sebastian Engell, Kai Sundmacher (Eds.), 2022
ISBN 978-3-11-070943-8, e-ISBN (PDF) 978-3-11-070985-8,
e-ISBN (EPUB) 978-3-11-070991-9

Vahid Pirouzfar, Yeganeh Eftekhari,
Chia-Hung Su

Pinch Technology

Energy Recycling in Oil, Gas, Petrochemical
and Industrial Processes

DE GRUYTER

Authors
Prof. Dr. Vahid Pirouzfar
Department of Chemical Engineering
Islamic Azad University
Central Tehran Branch
Ashrafi Esfahan Highway
Tehran
Iran
v.pirouzfar@iauctb.ac.ir

Yeganeh Eftekhari
Department of Chemical Engineering
Islamic Azad University
Central Tehran Branch
Ashrafi Esfahan Highway
Tehran
Iran
Yeganeh.eftekhary@yahoo.com

Chia-Hung Su
Department of Chemical Engineering
Ming Chi University of Technology
Gongzhuan Road
Taishan, New Taipei City
Taiwan
chsu@mail.mcut.edu.tw

ISBN 978-3-11-078631-6
e-ISBN (PDF) 978-3-11-078632-3
e-ISBN (EPUB) 978-3-11-078643-9

Library of Congress Control Number: 2022931840

Bibliographic information published by the Deutsche Nationalbibliothek
The Deutsche Nationalbibliothek lists this publication in the Deutsche Nationalbibliografie;
detailed bibliographic data are available on the Internet at http://dnb.dnb.de.

© 2022 Walter de Gruyter GmbH, Berlin/Boston
Cover image: Formosa Plastics Group
Typesetting: Integra Software Services Pvt. Ltd
Printing and binding: CPI books GmbH, Leck

www.degruyter.com

Preface

This book is presenting to explain the principles of process integration, the use of pinch technology as well as energy recycling in oil, gas, petrochemical and industrial processes.

We made every effort to prepare this book as a complete content overview of all aspects in similar references in the field of energy recovery in oil, gas and petrochemicals, by considering standards and the way of performing heat exchangers network and implementing the engineering, procurement and construction projects also the helpful experiences in optimizing them via latest process modeling of the largest engineering companies, for those who are involving in this field and would be focusing on their future plans.

It is hoped that this book will serve as a guide and reference for professors, technicians, craftsmen, managers, students and graduated people in various engineering majors, especially chemical and process engineering. Meanwhile the content presented provides enough basic information as feed for all students interested in energy industry and its technical, process and operational topics. We wish that the contents of this book would be a guide for developing the future and various applicable plans for the energy, oil, gas and downstream industries.

In case you have any criticisms, suggestions or solutions for improving this book and other similar books that are being written, it would be our pleasure if you contact us through the aforementioned communication channels.

https://doi.org/10.1515/9783110786323-202

Contents

Foreword

In the past few decades, by developing technology and industry day by day, due to the decrease in energy resources, the integration and optimization of energy consumption have become one of the main programs and priorities of industrial units, for instance, petrochemicals and refineries. In these units, a high level of energy usage in addition to imposing heavy costs would increase the environmental pollutants, which the sponsors and investigators of environmental organizations have always opposed. Despite the improvement and development of various industries and many benefits, the excessive consumption of raw materials and energy produces toxic and polluted substances, causing irreparable damages to the environment.

Therefore, energy optimization and increasing the rate of recovering energy have been always the most important issues in modification and design of industrial processes. Energy recovery systems and methods will prevent the losses in a process and would utilize the wasted energy to heat or preheat the process which is mandatory. Therefore, the energy that has been lost would be removed from the process or might be converted into a helpful energy for the process. Hence, the first task in heat recycling analysis is to discover the energy losses which can be recovered in a process and to consider the process unit from the sight of energy. In this regard, with the aim of process integration and optimization of energy consumption, pinch technology is a widely used and applicable way to design and optimize many of the processes and industrial units.

The most important applications of pinch technology are as follows:
– Design and optimization of heat exchangers network
– Proper and optimal selection of hot and cold utilities
– Heat integration and optimization of distillation towers, reactors and pumps
– Reducing the generation of pollutant gas

In this book, by presenting pinch technology and investigating on that, we attempt to describe and clarify the fundamental concepts to gain composite curve, grid diagram and optimal process design. May the engineers who are interested in the field of process design and optimization find this guidebook as one of the most helpful references.

https://doi.org/10.1515/9783110786323-204

About the Authors

Vahid Pirouzfar earned his PhD in chemical process engineering, process design, simulation and control from Tarbiat Modares University. He has been working as a faculty member at the Islamic Azad University (IAU) since 2011, and currently he is a faculty member and associate professor at IAU of Central Tehran Branch.

Since 2007 he has been working in large companies, refineries, consultant and contractor research institutes such as Oil and Energy Industries Development Company, Central Petrochemicals (four big petrochemical complexes producing ethylene, high-density polyethylene, low-density polyethylene (LDPE) and linear LDPE), Energy Renovation and Reclamation Consultants (Mabna), Iran Industrial Consultants Company, Middle East Energy Development Engineers Company, and also the Oil and Energy Research Institutes in the fields of oil, gas, petrochemical and energy in upstream and downstream industries in various occupations as expert, master, supervisor, head of the project and engineering department and managing director. He cooperated with internationally reputed companies like Shell and Sinopec during his activity.

Dr. Pirouzfar is a writer of numerous international articles and books. He has proposed many different presentations and reports in the field of engineering. He is also the editor, referee, manager and scientific executive director of many international and national journals and conferences. In addition to teaching various university courses, he has organized so many managerial and engineering courses, especially the process design and simulation engineering specialized courses. His favorite fields of training and researches include conceptual design of industrial processes, process integration, pinch technology, separation and membrane processes, combustion safety, modeling and simulation, and fuel and energy. He was the top graduate of Tarbiat Modares University at the doctorate level and the Iranians brilliant talent from the Ministry of Science.

V.Pirouzfar (*PhD; Control, Simulation and Design Of Chemical Processes Department*)
Department of Chemical Engineering, Islamic Azad University, Central Tehran Branch

P.O. Box: 14676–8683, Tehran, Iran
Tel: +98-912-2436110
Email: v.pirouzfar@iauctb.ac.ir
Google scholar: https://scholar.google.com/citations?user=ktSOIHgAAAAJ&hl=en
Research Gate: https://www.researchgate.net/profile/Vahid_Pirouzfar
ORCID: http://orcid.org/0000-0002-2862-008X
Linkedin: https://www.linkedin.com/in/vahid-pirouzfar-0273b137
HomePage: http://v-pirouzfar-chemeng.iauctb.ac.ir/faculty/en

Google scholar page: ORCID: Personal page:

https://doi.org/10.1515/9783110786323-205

Yeganeh Eftekhari earned her master's degree in chemical process engineering from Islamic Azad University of Central Tehran Branch and her BSc in petroleum engineering specialized in process design. She has been working as an expert since 2014 in oil, gas, petrochemical and engineering, procurement and construction contractor companies. She has proposed presentations, reports and researches in the field of process simulation, optimization of energy consumption as well as pinch analysis.

Y. Eftekhari (*MSc; Chemical Processes Engineer*)
Tel: +98-910-9281302
Email: Yeganeh.eftekhary@yahoo.com
          Yeganeheftekhari6@gmail.com

Chia-Hung Su received his PhD in chemical engineering from National Tsing Hua University in 2007. In 2009, he joined Ming Chi University of Technology where he is now professor and head of the Department of Chemical Engineering. Dr. Su has published over 100 technical papers, and his teaching and research interests are in the areas of process systems and control engineering for microbial, biochemical and complex chemical processes.

Chia-Hung Su (*PhD*)
Department of Chemical Engineering, Ming Chi University of Technology
Tel: +886 933 732 616
Email: chsu@mail.mcut.edu.tw

Chapter 1
Overview on the History of Integration and Fundamental Concept

1.1 Overview on the History of Pinch Technology

Reducing the energy consumption and industrial unit expenses is the factor that has been always studied and researched in process optimization issues. Generally, the results are showing that by applying changes and providing suitable operational conditions by an accurate process control and monitoring, using standard methods and engineering design could help achieve this purpose.

In this regard, many concepts have been spread and process integration is relatively a new concept and widely used today in industries, in order to examine a particular part of process design.

1.1.1 Process Integration

A chemical and industrial process consists of a set of related units and many different streams. Process integration is a general way to design and run a process, and actually, it emphasizes on integrity of the process, which includes several parts.

Due to the evolution that occurred in process integration during recent decade, the International Energy Agency has presented a definition of process integration as follows:

> In fact, process integration is the application of methods and algorithms which are based on system and through a comprehensive approach, it can lead to new design, modification and integration of industrial processes. These methods might be mathematical, thermodynamic and economic models and algorithms, that the Artificial intelligence, pinch analysis and mathematical programing are also some examples of them.

In other words, the main purpose of process integration is an integrated system analysis in such a way that the interaction of components in system is examined comprehensively in order to improve the design aims or utilization [1].

Process integration includes the following items:
- Optimization of energy consumption
- Optimization of external heat sources system and utilities
- Improving the process and operational conditions of equipment such as distillation towers and reactors
- Reducing operational and capital costs
- Optimization of hydrogen consumption

https://doi.org/10.1515/9783110786323-001

- Reducing greenhouse gas emissions
- Optimization of sequential separation processes and discontinuous processes
- Reducing the rate of water consumption and effluent production

Heat recovery system is one of the basic parts of process units and has a decisive and key role to play in energy consumption of units, determining heat cost and promoting heat recovery in chemical processes. This system includes heat exchanger network, cold utilities and hot utilities.

Therefore, using process integration methods in refinery units such as petrochemical units has been developed and expanded. In this regard, many new methods have been invented, and the most important ones are pinch analysis and mathematical programming method.

1.1.2 Mathematical Programming Method

For so long, the energy supplying system of the process heat network was performed traditionally. In this method, the process core is firstly designed by constant flow rates and temperatures, and the mass and energy balances of system take place. After that the design of heat recovery system would be completed. At the end, other system requirements will be covered by the use of external heat sources.

Due to the high energy consumption and operational costs in classical methods, researchers by presenting mathematical methods have provided solutions in order to optimize energy and operating costs of industrial units.

In 1969, Kesler and Parker proposed a linear programming based on a simultaneous algorithm. In this method, which aims to minimize the total costs of process heat network, each stream splits into multiple and smaller streams that are equal and have specific heating load. Then the smaller hot flows will be connected to the cold flows. To expand more, networks that include all substreams must be built.

In 1973, sequential algorithms method and tree searching algorithm have been presented by Lapidus, which eventually created a tree diagram to design a heat network. But the disadvantage of this method is that it could only be applied on the systems with less than ten streams.

However, the most significant researches which have been done through mathematical method are based on the researches by a group of scholars such as Floudas and Grossmann: By using a nonlinear model, Floudas and his colleagues have studied various structural changes in the network at the same time, including changing the position of exchangers, reusing piping and adding new exchangers to heat network.

Grossmann and his teammates presented the mixed integer nonlinear programming for targeting and design of the network, which caused improvement of accuracy and efficiency of the model. But due to an increase in calculation of volume, the cost and time had grown as exponential functions of the exchangers.

As most of the proposed models are based on mathematical models, there would be still two main problems:

- First, because of using extensive structures, math problems become so large, then the problem solving gets impossibly hard and takes lots of time and cost.
- Second, since the target function usually has local optimal points, the obtained answer may not be the absolute optimal response.

1.2 Introduction of Pinch Technology

Due to the weakness of the theoretical foundations and algorithms of solving optimization problems and methods in the past, using mathematical methods has always been limited. Therefore, the researches focused on identifying the optimization aims, based on optimizing energy consumption and economic analysis of exchanger's network and heat sources. In this regard, the idea of many researchers and scholars turned to creation of pinch technology [2].

The advent of pinch technology can be considered as a method of heat recovery at the pinch point which was invented separately by scientists such as Hohmann, Linnhoff and Umeda in the 1970s.

Since the pinch technology has a wide range of applications with high efficiencies in optimizing industrial units, it has been developed and more completed through lots of efforts and researches over the years.

Along with the chapters in this book, by investigating the pinch technology, we intent to provide some methods in order to optimize heat exchangers network and flow diagrams, as well as minimizing energy consumption of industrial units.

Firstly, Figure 1.1 shows a schematic of flowsheet for an industrial unit, which is representing a traditional design. Six heat transfer units including heaters, coolers and exchangers are used, and the energy requirements are 1,722 kW for heating and 654 kW for cooling. Figure 1.2 shows an alternative design using pinch analysis technology for optimal design with energy targeting and heat exchanger network integration.

Therefore, in alternated flow diagram, only four heat exchangers are used, and the external source of cooling water has been removed from the process unit. Furthermore, in alternative design, the heat load of the heating source has been decreased from 1,722 to 1,068 kW, which causes a 40% reduction in total energy of the process unit.

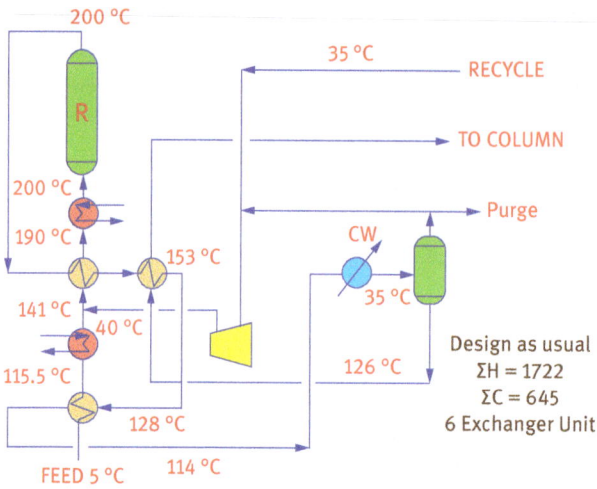

Figure 1.1: Traditional process design and heat network of an industrial unit.

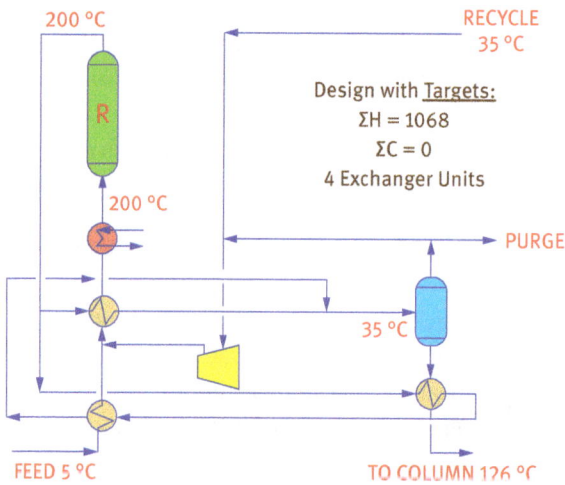

Figure 1.2: Process design of an industrial unit by using pinch technology.

In the schematic expression of an industrial unit, it can be well recognized that in industrial factories, there are so many opportunities to save energy, which could optimize these units by identifying and providing solutions.

1.3 Basic Definitions

Before analyzing the pinch technology, we would introduce the basic concepts and definitions which are required to be known in pinch technology.

1.3.1 The Process Stream

The streams in industrial units can be divided into two categories: hot streams and cold streams.

Hot streams: These flows will move from a high temperature to a lower temperature by losing energy, for instance, the exhaust steam from the top of distillation tower:

$$T_s \rightarrow T_t, \ T_s > T_t$$

T_s is the supply temperature and T_t is the target temperature.

Cold streams: These flows receive energy and rise from a lower temperature to a higher temperature, for example, the drained outlet flow from the bottom of a distillation tower:

$$T_s \rightarrow T_t, \ T_s < T_t$$

1.3.2 Utilities

In order to fully supply the heat requirements in industrial units, we must use utilities, which are divided into two categories: hot utility and cold utility.

Hot utility: This is the source that provides heat to the unit but is not considered as process flow such as furnace and steam utility (at different temperature levels).

Hot utilities and hot streams are the same in terms of supplying energy, but hot utility is not considered as process flow; however, the hot stream is a part of process flow [3].

Heaters: These are the equipment in which cold stream is a part of process flow and their hot stream is part of hot utility.

Cold utility: This is the source that supplies cooling to the unit but is not included in process flows, such as refrigerant utility (in different temperature levels), cooling water utility and air cold utility.

Cold utilities and cold streams are the same in terms of receiving energy, but cold utility is not considered as process flow; however, the cold stream is a part of process flow [4].

Coolers: These are the equipment in which hot stream is a part of process flow and their cold stream is part of cold utility.

Heat exchangers: These are equipment that exchange the heat and in which their hot and cold streams are considered as process flow.

One of the main aims in integration of process and heat exchangers network is to increase the heat load of exchangers and decreasing the heat load of hot and cold utilities. In other words, increasing energy recovery and reducing the use of utilities are favorable goals in process integration.

1.3.3 Sensible Heat

Sensible heat of each flow represents the amount of energy that the flow losses or receives. This heat can be calculated through the following equation [5]:

$$Q = FC_P(T_s - T_t)$$

where Q represents sensible heat, F is the mass flow rate, C_P is the specific heat capacity, T_s is the flow supply temperature, T_t is the flow target temperature.

If we introduce the result for multiplication of specific heat capacity and mass flow rate as the flow heat capacity and denote it by C_P, the above equation changes as follows:

$$Q = C_P(T_s - T_t)$$

Note that while the phase is changing, the sensible heat will be calculated by the following equation:

$$Q = F\lambda$$

In this equation, λ is the latent heat.

Example 1.1: Information about the two streams in the process is given below. Determine how the energy of these two flows will be supplied?

No. of flow	Flow type	Supply temperature (°C)	Target temperature (°C)	Heat capacity (MW/°C)
1	Hot	150	50	0.1
2	Cold	50	100	0.2

Answer: Flow no. 1: It is considered as hot stream because it must raise from a higher temperature to a lower temperature. The sensible heat of flow no. 1 is calculated as follows:

$$Q_1 = C_P(T_s - T_t) = 0.1(150 - 50) = 10 \text{ MW}$$

It means that this flow must lose 10 MW of energy to reach 50 °C from the temperature of 150 °C.

Flow no. 2: It is considered as cold flow because it must rise from a lower temperature to a higher temperature. The sensible heat of this flow is calculated as follows:

$$Q_2 = C_P(T_s - T_t) = 0.2(50 - 100) = -10 \text{ MW}$$

It means that this flow must receive 10 MW of energy to reach the temperature of 100 °C from 50 °C. The following methods can be used to supply required energy by these two streams:

First Method: Using Utilities
In this method, hot stream transmits 10 MW of energy to cold utility and cold stream receives 10 MV of heat energy from hot utility.

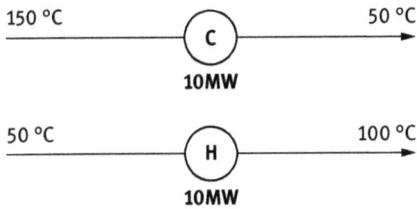

150 °C ——————(C)—————→ 50 °C
                10MW

50 °C ——————(H)—————→ 100 °C
               10MW

As it is obvious, in this method, the energies in the process have not been used at all and only hot utilities have been used to supply energy. Therefore, this method does not seem to be a desired and suitable method for designing the exchanger network in the process unit.

Second Method: Using Heat Exchanger
In this method, the hot and cold streams in the process are entered into a heat exchanger. The hot stream loses 100 MW of energy and reaches the temperature of 150–50 °C, while the cold stream is receiving 10 MW of energy from the hot stream and rises from the temperature of 50–100 °C.

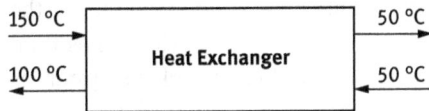

150 °C →| | 50 °C
        | Heat Exchanger |
100 °C ←| | 50 °C

In the last sections, we explained that in the process of heat integration, increasing the heat load of the exchanger and reducing the heat load of utilities are the main goals in designing the heat exchanger network of process units.

As it is clear, in this method, all energy in the process has been used but no utility has been consumed.

Now the question is whether this method is the desired method or not.

The answer to this question is no. Because on one side of the exchanger, ΔT is equaled to zero. Therefore, according to the following equation, the required area of heat transferring in the exchanger will tend to be infinite:

$$Q = UA\Delta T$$

where Q is the exchanger heat load, U is the exchanger heat transfer coefficient, ΔT is the temperature difference, and A is the exchanger area.

Therefore, we conclude that although in this method, we have reduced the consumption of utility, but on the other hand, we have increased the exchangers surface area to infinity. Though we have reduced the energy cost, the equipment cost will be infinite. As a result, we recognize that this method is not favorably the same as the first method.

To solve this issue, we firstly express the concepts of first and second laws of thermodynamics.

1.3.4 Concept of the First Law of Thermodynamics

The first law of thermodynamics, also known as the law of conservation of work and energy, represents the macroscopic equilibrium state of a system and is expressed by a factor called potential energy [6].

If the system interacts with the environment, it changes from its primary macroscopic state to its secondary macroscopic state. In this process, a change in the potential energy of the system is shown as follows:

$$\Delta U = Q + W \rightarrow \Delta U = F \cdot C_P \cdot \Delta T + P \cdot \Delta V$$

where W represents the macroscopic work done by the system and Q represents the amount of heat transferred during the process.

In a process that will run with a constant pressure, the sum of the work done and the heat transferred from the system to the environment is equal to the enthalpy difference of the system:

$$\Delta U = \Delta H = Q + W$$

In constant volume processes, where the fluid volume does not change during the process, the work done is zero. Therefore, in these types of processes such as fluid heat transferring inside the heat exchanger, the potential energy of the system is equal to the heat absorbed by the system or the heat transfer rate from the system to the environment:

$$\Delta U = F \cdot C_P \cdot \Delta T + P \cdot \Delta V, \quad \Delta V = 0 \rightarrow \Delta U = Q = F \cdot C_P \cdot \Delta T$$

In constant pressure and volume processes, the enthalpy change will be equal to the rate of heat transfer of the system or the sensible heat:

$$\Delta U = \Delta H = Q + W, \quad \Delta V = 0 \rightarrow \Delta H = F \cdot C_P \cdot \Delta T$$

1.3.5 Concept of the Second Law of Thermodynamics

Heat is transferred from a higher temperature to a lower temperature. In other words, we would need a driving force to transfer the heat, and the temperature difference is the driving force for heat transfer [7, 8].

In actual processes, both the first and second laws of thermodynamics must be considered. So when we put two hot and cold flows into a heat exchanger, we have to follow both the laws of thermodynamics.

Now we return to Solution 1.1. Note the exchanger which we have used in the second method:

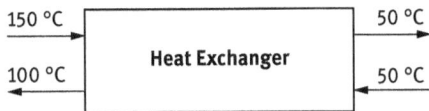

```
150 °C                                     50 °C
    ┌──────────────────────────┐
    │      Heat Exchanger      │
100 °C                             50 °C
    └──────────────────────────┘
```

Hot flow loses 10 MW of energy. If we ignore the amount that has been lost, the cold flow receives a total of 10 MW from the hot flow. Therefore, the first law of thermodynamics is true and applied.

Now pay attention to the temperature difference between the left and right sides of the exchanger. The temperature difference on the left side of the heat exchanger is 50 °C, while on the right side of it the temperature difference is zero, which has broken the second law of thermodynamics.

Hence, in heat exchangers, a parameter named ΔT_{min} will be defined.

1.3.6 Minimum Allowable Temperature Difference in Heat Exchangers (ΔT_{min})

ΔT_{min} refers to the minimum allowable temperature difference in heat exchangers. As an example if $\Delta T_{min} = 10$ °C, the temperature differences must not be less than this amount at any part of the exchanger (but it is permitted to be more than this amount).

The parameter "ΔT_{min}" plays an important decisive role in the area of exchangers and the consumption of utilities. The increase in ΔT_{min} value is defined/considered as follows:
1. Increasing the heat load of utilities aims to increase the energy costs and the reduction in energy recycling.
2. Decreasing the surface area of heat exchangers reduces the capital costs and expenses for the equipment and devices.

The decrease in ΔT_{min} is defined as follows:
1. Decreasing the heating load of utilities aims to reduce the energy costs and increase the energy recycling.
2. Increasing the surface area of heat exchangers increases the capital costs and expenses for the equipment and devices.

With reference to the above description, ΔT_{min} is a function of energy costs and capital cost (Figure 1.3). Therefore, as the total cost is evaluated at a lower level, the value of ΔT_{min} would be optimal.

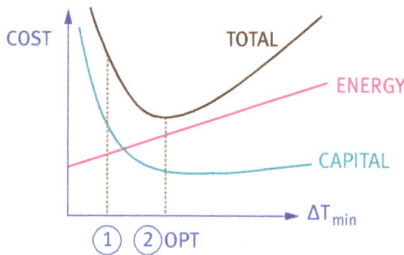

Figure 1.3: Diagram of industrial unit cost in comparison to ΔT_{min}.

This diagram represents an important point that the process of heat integration is not defined by minimizing the energy consumption but actually the optimization of that. As it is shown in the above diagram, while the energy is at the minimum level of itself, equipment cost becomes infinite.

It should be noted that since the energy and equipment costs vary in different countries, for a specific project, the value for optimal ΔT_{min} differs from one country to another.

Now in Example 1.1, we assume that ΔT_{min} is equal to 10 °C. In this case, the hot stream must come out from the exchanger with the temperature of 60 °C at least, so that the temperature difference between the hot and cold inlet flows reaches to 10 °C, and the sensible heat of the hot stream in the exchanger will be calculated as follows:

$$Q = 0.1\,(150 - 60) = 9\,\text{MW}$$

In this regard, hot stream loses 9 MW energy inside the exchanger. This amount of heat is received by the cold stream. As a result, according to the sensible heat equation, the temperature of outlet cold stream from the exchanger can be calculated as follows:

$$-9 = 0.2\,(50 - T_t) \rightarrow T_t = 95\,°C$$

Now the hot stream will go through the cooler in order to reach the temperature of 50 °C, and cold stream will get into the heater to catch the temperature of 100 °C.

The heating load of heater and cooling load of cooler can be calculated from the following equations:

$$\text{Cooling load of cooler} \rightarrow Q_C = 0.1(50 - 60) = 1 \text{ MW}$$

$$\text{Heating load of heater} \rightarrow Q_H = 0.2(100 - 95) = 1 \text{ MW}$$

Therefore, the schematic of heat exchanger of the process in Example 1.1 is as follows:

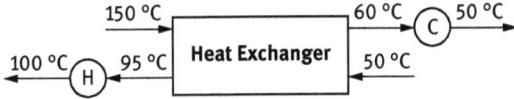

As a result, by using the applied method, we have been able to get the most out of the energy of the process streams in order to exchange the heat.

Chapter 2
Composite Curve

In the last chapter we have been learning the basic concepts of process integration and now we present the pinch analysis.

Many different examples are given in each section for better understanding of the concepts; by solving them, we intend to explain and present the pinch topics as well as possible.

2.1 Data Extraction

The first step in pinch analysis is data extraction, which stands for identifying the process hot and cold streams and extracting all data belonging them into a table, such as supply and target temperatures, heat capacity and sensible heat (enthalpy difference).

Example 2.1: Please consider the process shown in Figure 2.1, then extract the process data and add them into a table.

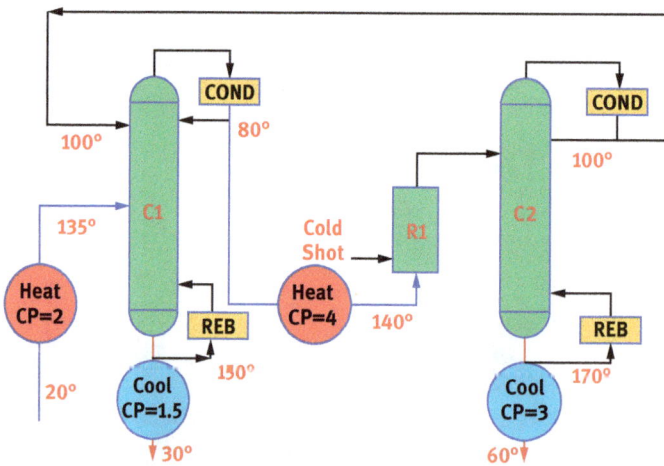

Figure 2.1: Process schematic of an industrial unit.

https://doi.org/10.1515/9783110786323-002

Answer: In this way, the data of the above process are summarized in the table as follows:

Stream no.	Stream type	Supply temperature (°C)	Target temperature (°C)	Heat capacity (MW/°C)
1	Hot	170	60	3
2	Hot	150	30	1.5
3	Cold	80	140	4
4	Cold	20	135	2

2.2 Composite Curve

In this diagram, vertical axis displays the temperature and horizontal axis displays the enthalpy (energy) difference [9]. The slope of this diagram is equal to $1/C_P$ according to the following equation:

$$\Delta H = F \cdot C_P \cdot \Delta T = C_P \cdot (T_s - T_t)$$

At first, we would study a simple case including a hot and a cold stream.

Example 2.2: The information regarding two hot and cold streams are presented in the following table. Draw the composite curve ($\Delta T_{min} = 10\ °C$).

Stream no.	Stream type	Supply temperature (°C)	Target temperature (°C)	Enthalpy (MW)
1	Cold	30	100	−14
2	Hot	150	30	12

Answer: Firstly, we draw the hot flow in a specific location on the diagram of enthalpy variation with temperature ($T - H$). Then draw the cold stream at the right of the hot stream and bring it closer to hot stream that at one point the distance becomes equal to ΔT_{min}. This point is named pinch point.

The area between B and C is where the two hot and cold streams can exchange heat with each other (according to Figure 2.2, in this area the two streams overlap). The amount of energy transferred in this area is shown as Q_{REC}. In accordance to the figure in this example, the value is 11 MW.

With reference to the table for the hot flow:

$$\Delta H = C_P \cdot (T_s - T_t) = C_P\,(150-30) = 12\ \text{MW} \implies C_{P_{Hot}} = 0.1\ \text{MW}/°C$$

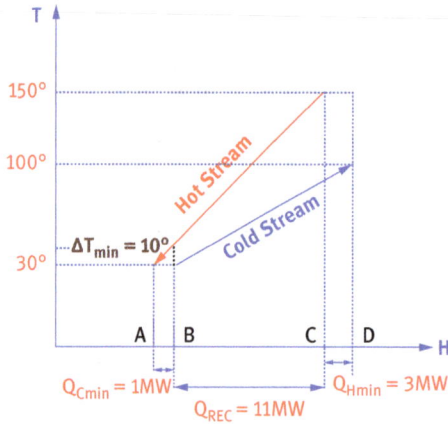

Figure 2.2: Composite curve of Example 2.2 ($\Delta T_{min} = 10$ °C).

By considering the diagram in the overlapping area between hot and cold streams, we will have

$$Q_{REC} = C_P \cdot (T_s - T_t) = 0.1 \, (150 - 40) = 11 \, \text{MW}$$

The area between A and B is where the hot stream must lose the remaining heat. This amount of heat must be transferred to a cold utility. Therefore, the energy of this area indicates the cooling load of the cooler, which is shown by Q_{Cmin}. In this example, according to the enthalpy of hot stream and the Q_{REQ} in the diagram, the value of Q_{Cmin} is equal to 1 MW.

Actually, Q_{Cmin} represents the minimum amount of energy that must be transferred to a cold utility in the process.

The area between D and C is where the cold stream must supply its remaining energy which is required, and this amount of energy must be given to cold stream by a hot utility. Therefore, the energy of this region indicates the heating load of the heater, which is shown by Q_{Hmin}. In this example, in accordance to the figure the value of Q_{Hmin} is equal to 3 MW.

Actually, Q_{Hmin} represents the minimum amount of energy that must be supplied from a hot utility.

Example 2.3: Solve Example 2.2 with $\Delta T_{min} = 20°C$.
According to the explanations given in Example 2.2, the composite curve for $\Delta T_{min} = 20$ is shown in Figure 2.3.

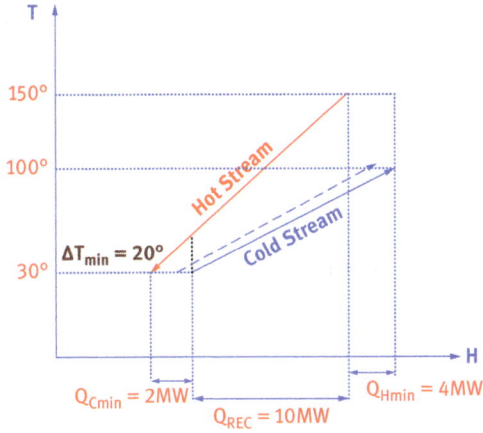

Figure 2.3: Composite curve of Example 2.3 (ΔTmin = 20 °C).

Considering these two examples, we would discover that if ΔT_{min} increases, the heat load of exchanger decreases, but the heat load of hot utility increases.

Example 2.4: Draw the composite curve for Example 2.1 ($\Delta T_{min} = 10°C$).

Answer: As you see, the number of streams are more than the last two examples (Figures 2.4 and 2.5), so how to draw a composite curve is bit more complicated. At first, we draw hot streams on one diagram and the cold streams on another.

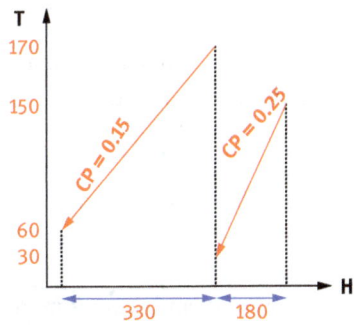

Figure 2.4: Composite curve for each hot stream regarding Example 2.4.

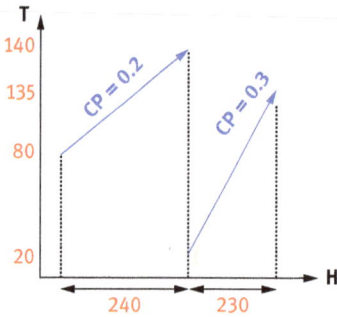

Figure 2.5: Composite curve for each cold stream regarding Example 2.4.

Now we get the sum of hot and cold streams separately, and then we will show them as two different streams.

In order to sum up the hot streams, we would do as follows:

1. At the temperature range of 150–170 °C, there are just the streams with heat capacity of $C_P = 3$ MW/°C. So the enthalpy difference in this region is equal to

$$\Delta H = 3 \ (170{-}150) = 60 \text{ MW}$$

2. At the temperature range of 60–170 °C, there are two streams with heat capacities of 3 and 1.5 MW/°C. So firstly, we get the sum of streams' heat capacities, and then we calculate the enthalpy difference in this region as follows:

$$\Delta H = 4.5(150{-}60) = 405 \text{ MW}$$

3. At the temperature range of 30–60 °C, there is only one stream with a heat capacity of 1.5 MW/°C. Therefore, the enthalpy difference in this region is equal to

$$\Delta H = 1.5(60{-}30) = 45 \text{ MW}$$

As a result, the composite curve of the sum of hot streams is drawn as in Figure 2.6.

For cold streams, we would do the same as hot streams. Therefore, the composite curve of the sum of cold streams is also drawn as in Figure 2.7.

While these two diagrams are combined, the application of them becomes more important. In this regard, firstly we draw the hot stream on T–H diagram as in the last example, and then the cold stream at the right side and bring it closer to the hot stream; hence, at one point the distance between two graphs becomes ΔT_{min}, which is actually named the pinch point and divides the process into above and below the pinch point.

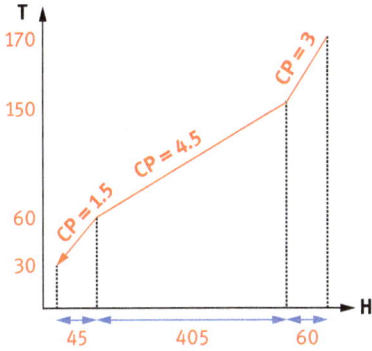

Figure 2.6: Composite curve of hot streams regarding Example 2.4.

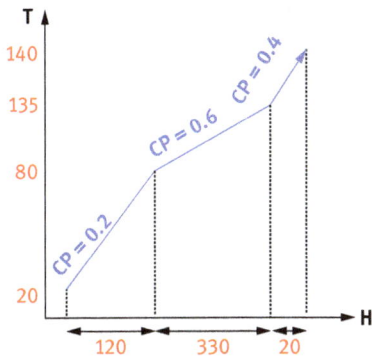

Figure 2.7: Composite curve of cold streams regarding Example 2.4.

Finally, regarding this process, the graph of $T-H$ will be drawn and Q_{Hmin} and Q_{Cmin} are estimated as follows (Figure 2.8):

$$\begin{cases} Q_{H_{min}} = 20 \text{ MW} \\ Q_{C_{min}} = 60 \text{ MW} \end{cases}$$

$$\text{Pinch point} = \begin{cases} 90 \,^{\circ}\text{C} & \text{Hot} \\ 80 \,^{\circ}\text{C} & \text{Cold} \end{cases}$$

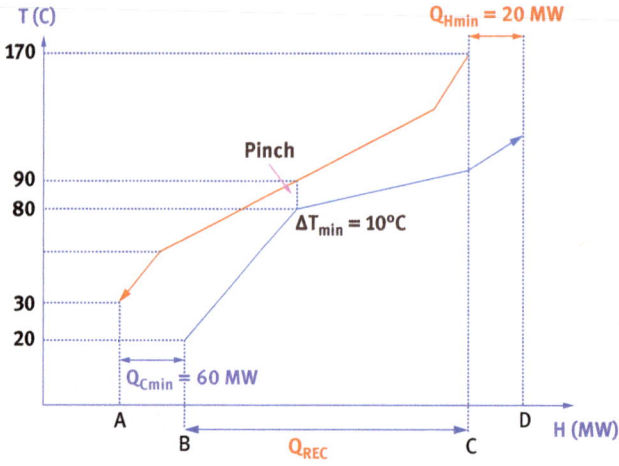

Figure 2.8: Composite curve of industrial unit regarding Example 2.4.

Chapter 3
Cascade Diagram

3.1 Golden Rules of Pinch Technology

In accordance with composite curve we can gain some important points which are known as pinch golden rules (Figure 3.1):

1. To design by considering minimum energy consumption, any of the energy flows are not allowed to pass through the pinch point.
2. Thermal energy must be added to the heat network only at the above of pinch point, Therefore, in order to design above the pinch point, only hot utility can be used to supply energy.
3. Thermal energy must be excreted from the heat network just below the pinch point. Therefore, in order to design below the pinch point, only cold utility can be used to receive excreted energy.

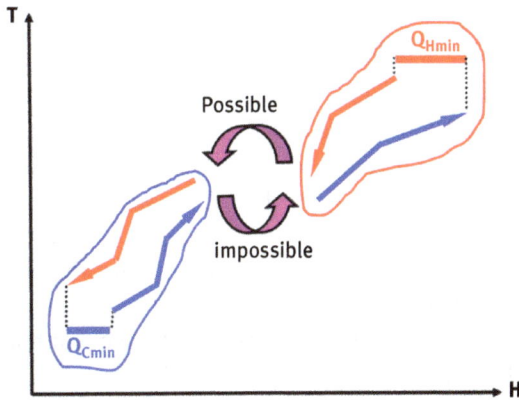

Figure 3.1: Composite curve for golden rules of pinch.

According to the $T–H$ diagram, points 2 and 3 are obvious but point 1 needs more description.

To explain this, consider that the amount of energy is transferred from above to the below point (Figure 3.2). So when this amount of energy is transferred from the top to the below of pinch, it causes the cold flow above the pinch to have a shortage of α-energy, in which this amount must be supplied by hot utility; therefore, the heating load of heater will increase by the value of α.

On the other hand, when this amount of energy is transferred below the pinch, it is an additional heat that must be excreted to cold utility. As a result, the cooling load of the cooler increases as much as α.

https://doi.org/10.1515/9783110786323-003

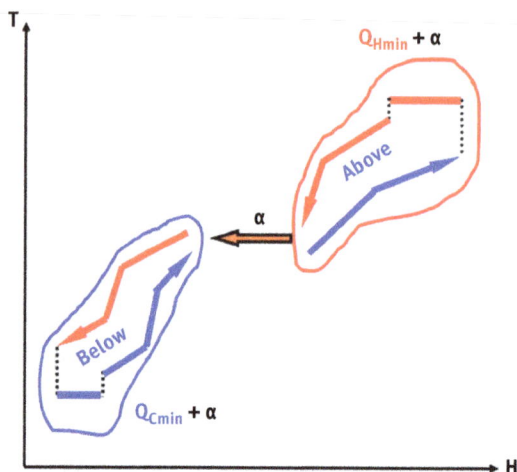

Figure 3.2: Composite curve of pinch, in case of heat transferring from above to below the pinch.

3.2 Cascade Diagram

In the last section, we studied how to specify the pinch point by using the $T–H$ diagram and to draw hot and cold streams [10]. Since there always might be some deviations and possibility of error in the drawing method, for solving this problem we shall introduce the cascade or cascade diagram. To draw this graph, we follow the instructions:

1. We consider two different axes for hot and cold streams. Then place the hot stream on the left and the cold stream on the right.
2. We determine the temperature corresponding to the temperatures of cold or hot streams according to the value of ΔT_{min}.

   For instance, assume that the temperatures of hot streams are 100 °C and ΔT_{min} is equal to 10 °C; in this regard, the temperature of the cold streams will be 90 °C.
3. We identify each flow according to the initial and final temperatures with a vertical vector on the diagram. Then we write the amount of heat capacity regarding each flow on the vector.
4. In every section, we calculate sensible heat through the following equation:

$$Q_i = \left[\sum (FC_P)_{Hot_i} - \sum (FC_P)_{cold_i}\right]\Delta T_i$$

5. In this diagram, the temperature decreases from top to down the graph. Therefore, in accordance to the second law of thermodynamics, heat is transferred from top to bottom. Whenever heat is required, it must be supplied by hot utility and at the end the energy of the last part below the pinch point shall be excreted to cold utility.

Note that the point at which there is no energy drop or transmitted is named the pinch point.

Example 3.1: Process data of two hot and cold streams are as follows, and by assuming $\Delta T_{min} = 10\ °C$, draw the cascade diagram.

No. of stream	Type of stream	Supply temperature (°C)	Target temperature (°C)	Heat capacity (MW/°C)
1	Hot	160	40	0.1
2	Cold	40	110	0.2

Answer: Firstly, we place the existing temperatures in two different axes and scales, and then by considering $\Delta T_{min} = 10\ °C$, we determine and write the temperatures opposite to the hot or cold stream. Furthermore, according to the supply and target temperatures of the streams, we indicate each stream via a vector and write the heat capacity on them:

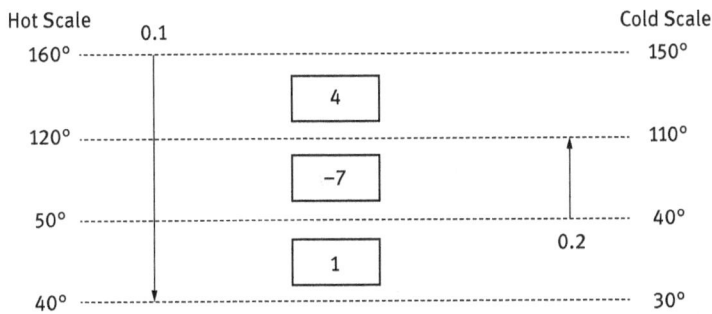

Figure 3.3: Cascade diagram of Example 3.1 (A).

As it is obvious in Figure 3.3, there are three regions that the sensible heat of each part is calculated as follows:
Sensible heat of the first region: $Q_1 = [0.1 - 0] \times (160 - 120) = 4\ MW$
Sensible heat of the second region: $Q_2 = [0.1 - 0.2] \times (120 - 50) = -7\ MW$
Sensible heat of the third region: $Q_3 = [0.1 - 0] \times (50 - 40) = 1\ MW$

With reference to the second law of thermodynamics, excess heat from the first zone can be transferred to the second zone. Therefore, the transmitting of 4 MW energy from the first zone to the second zone supplies the amount of heat demanded by this zone, but to meet the remaining energy we have to use hot utility. So we supply 3 MW of energy from a hot utility to this region (Figure 3.4). In this regard, no energy is conducted from the second area to the third area. At this point where there is no energy drop or transmission, it is named the pinch point. Finally, the third zone has 1 MW of energy that must be excreted to cold utility.

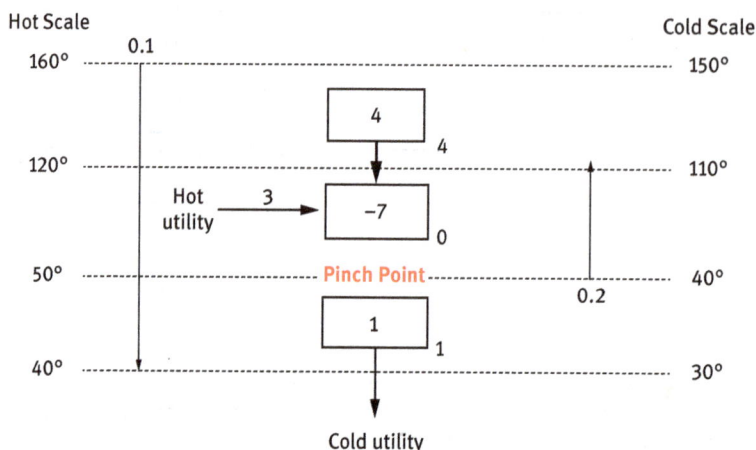

Figure 3.4: Cascade diagram of Example 3.1 (B).

Hence, the cascade diagram is completed and according to Figure 3.4, the pinch points, Q_{H_min} and Q_{C_min} are equal to

$$\text{Pinch point} = \begin{cases} 50\,°C & \text{Hot} \\ 40\,°C & \text{Cold} \end{cases}$$

$$Q_{H_{min}} = 3\text{ MW}, \quad Q_{C_{min}} = 1\text{ MW}$$

Example 3.2: Draw the cascade diagram of Example 2.1 and determine the pinch point accordingly.
Answer: As it is mentioned earlier in Example 3.1, first of all, we place the existing temperatures in two different axes and scales, then by considering $\Delta T_{min} = 10$ °C, we determine and write the temperatures opposite the hot or cold stream. We indicate each stream via a vector and write the heat capacity on them according to the supply and target temperatures of them (Figure 3.5).

Now we calculate the sensible heat for each zone:

$$Q_1 = [3 - 0] \times (170 - 150) = 60$$

$$Q_2 = [(3 + 1.5) - 4] \times (150 - 145) = 2.5$$

$$Q_3 = [(3 + 1.5) - (4 + 2)] \times (145 - 90) = -82.5$$

$$Q_4 = [(3 + 1.5) - 2] \times (90 - 60) = 75$$

$$Q_5 = [1.5 - 2] \times (60 - 30) = -15$$

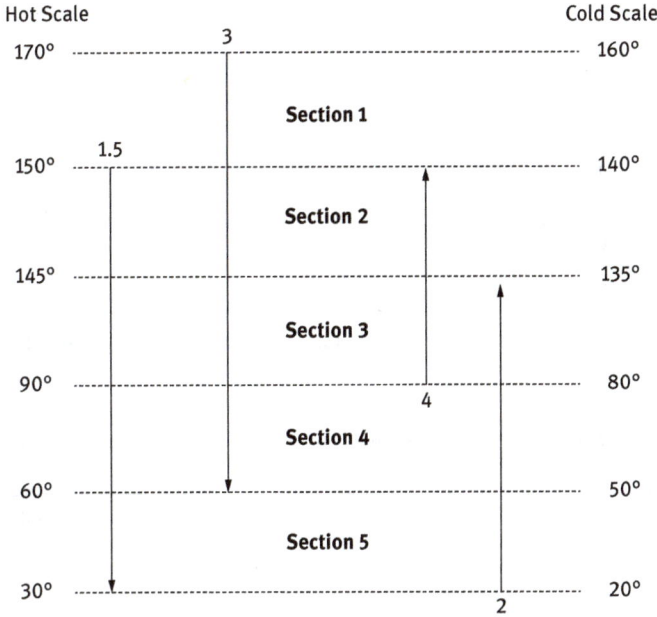

Figure 3.5: Cascade diagram of streams in Example 3.2 (A).

Afterward, we show the obtained values on the diagram and then conduct the energy from top to the down of cascade diagram graph (Figure 3.6).

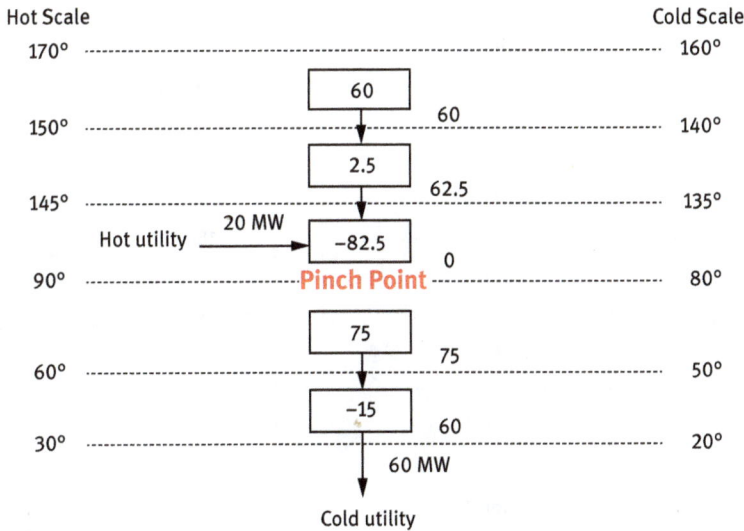

Figure 3.6: Cascade diagram of Example 3.2 (B).

Therefore, the pinch points $Q_{(H\_min)}$ and $Q_{(C\_min)}$ are equal to

$$\text{Pinch point} = \begin{cases} 90\,°C & \text{Hot} \\ 80\,°C & \text{Cold} \end{cases}$$

$$Q_{H_{min}} = 20\,MW, \quad Q_{C_{min}} = 60\,MW$$

As you see, the obtained values from the cascade diagram are similar with the values which are gained from the composite curve.

Example 3.3: With reference to the information in the below table, draw the cascade diagram for this process ($\Delta T_{min} = 10\,°C$).

No. of stream	Type of stream	Supply temperature (°C)	Target temperature (°C)	Heat capacity (MW/°C)
1	Cold	20	180	0.2
2	Cold	140	230	0.3
3	Hot	250	40	0.15
4	Hot	200	80	0.25

Answer: Firstly, we draw the basic cascade diagram and then we calculate the sensible heat of each area (Figure 3.7).

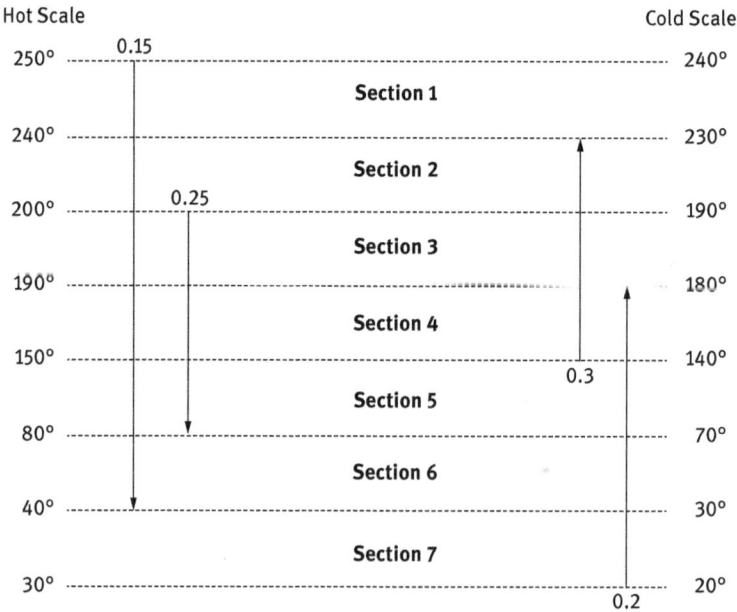

Figure 3.7: Cascade diagram of streams in Example 3.3 (A).

$$Q_1 = [0.15 - 0] \times (250 - 240) = 1.5 \text{ MW}$$

$$Q_2 = [0.15 - 0.3] \times (240 - 200) = -6 \text{ MW}$$

$$Q_3 = [0.15 + 0.25 + 0.3] \times (200 - 190) = 1 \text{ MW}$$

$$Q_4 = [(0.25 + 0.15) - (0.3 + 0.2)] \times (190 - 150) = -4 \text{ MW}$$

$$Q_5 = [(0.25 + 0.15) - 0.2] \times (150 - 80) = 14 \text{ MW}$$

$$Q_6 = [0.15 - 0.2] \times (80 - 40) = -2 \text{ MW}$$

$$Q_7 = [0 - 0.2] \times (40 - 30) = -2 \text{ MW}$$

Now we indicate the obtained values on the diagram and then we conduct energy from top to down the graph (Figure 3.9):

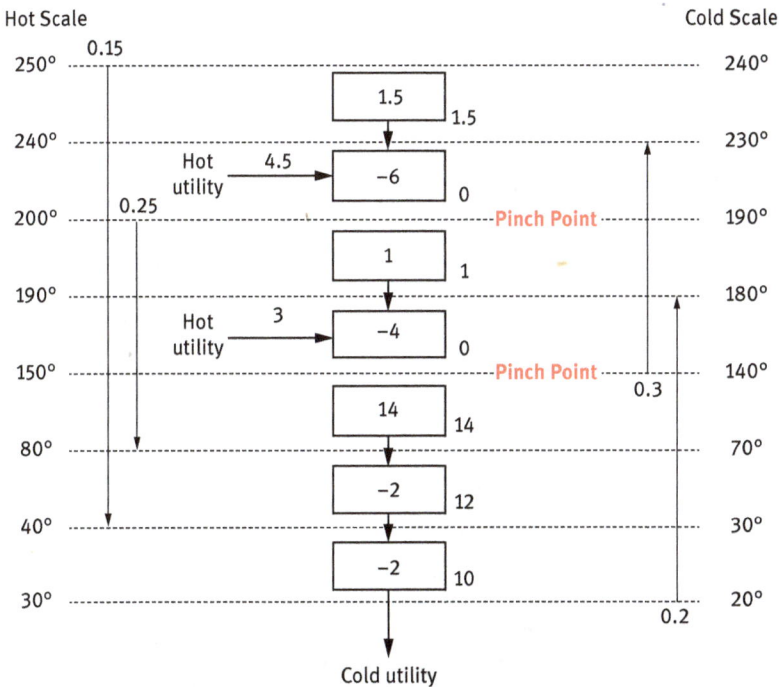

Figure 3.8: Cascade diagram of Example 3.3 (B).

As it is clear in Figure 3.8, the first area transfers 1.5 MW of energy to the second area. So the second area needs 4.5 MW of energy that must be supplied by a hot utility; therefore, the final part of the second area will be the pinch point.

The third area will transmit 1 MW of its energy to the fourth area, and the fourth area shall supply the required energy from the hot utility. Hence, the final part in this region would not have any energy drop or transmission; definitely, this cannot be the pinch point. At this situation and when there are more than one pinch point we must proceed with completing the operation to the end of process. Then we summarize the whole received energy from the hot utility and we inject it to the process from the top of the graph to the first region and we shall follow the instructions once again (Figure 3.9). In the given example, the sum of energy from hot utility is equal to $4.5 + 3 = 7.5$ MW.

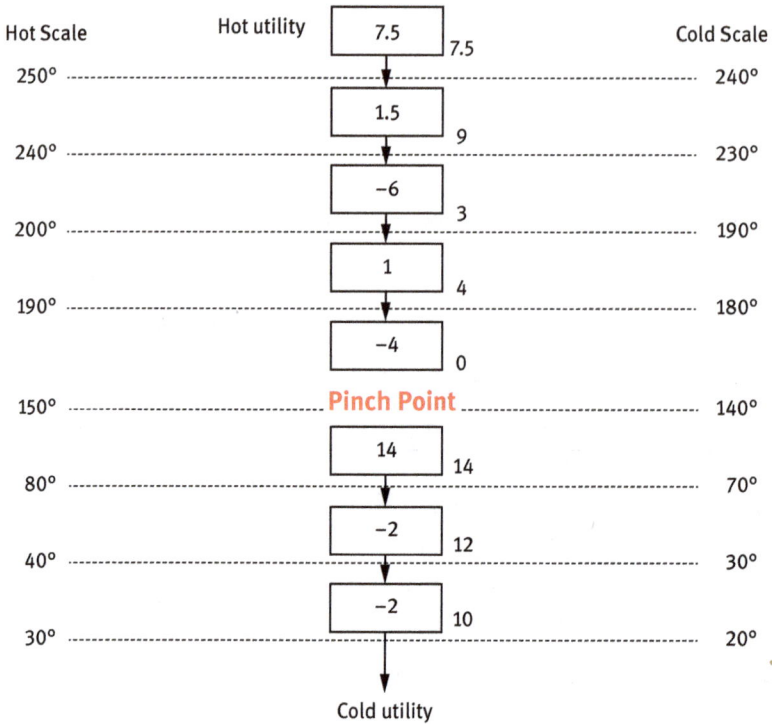

Figure 3.9: Cascade diagram of Example 3.3 (C).

Therefore, the pinch points $Q_{(H\_min)}$ and $Q_{(C\_min)}$ are equal to

$$\text{Pinch point} = \begin{cases} 150\,°C & \text{Hot} \\ 140\,°C & \text{Cold} \end{cases}$$

$$Q_{H_{min}} = 7.5\ \text{MW},\quad Q_{C_{min}} = 10\ \text{MW}$$

Chapter 4
Heat Exchanger Network Design with Maximum Energy Recovery

The most important part in pinch analysis and heat exchanger network design of a process unit is to recognize the design targets. Design targets or design objectives are the significant part of energy consumption monitoring plans in industrial projects. Generally, the aims of heat exchanger network design include the following two methods:

1. Design method with maximum energy recovery (MER)
2. Design method with minimum number of heat transfer units (N_{min})

In this chapter, we present the design method with MER, and in the next chapter, we examine the design method with minimum number of heat transfer units (N_{min}).

4.1 Heat Exchanger Network Design with Maximum Rate of Energy Recovery (MER)

In heat exchangers network design with a maximum rate of energy recovery, the aims are to make a maximum heat exchange between hot and cold streams in order to maximize energy recovery.

By designing with MER, we will see a reduction in energy costs and an increase in capital costs. So for heat exchangers network design, we shall use a grid diagram, and before investigation on the design methods we introduce the grid diagram.

4.2 Grid Diagram

The composite curve and grid diagram are not supposed to provide information for the design of heat exchanger's network and in order to achieve such information we shall use the grid diagram. To draw this graph, we stick to the following instructions:

1. Firstly, we draw a vertical dotted line and place the pinch hot temperature on above and the cold temperature below of it.
2. We draw hot streams from left to right and cold temperature from right to left.
3. We indicate heat capacities of streams at the right of the graph.

In this regard, the initial shape of the graph is prepared. Now we shall start designing the heat exchanger network. As explained earlier in the last sections, the pinch point thermally divides the whole process into two independent parts:

https://doi.org/10.1515/9783110786323-004

1. The upper side of pinch point, which has a higher temperature than the pinch point.
2. The lower side of pinch point, which has a lower temperature than the pinch point.

These two parts are different and independent. In other words, we have to design the upper and lower sides of the pinch point separately. At the end, by considering and combining the design of both regions, the design of the whole unit will be determined.

Regarding the heat exchanger network design, there are some rules that we would express them as discussed further.

4.3 Rules of Exchanger Design Above and Below the Pinch Point

1. Rule for the number of inlet and outlet streams to the pinch point is $N_{out} \geq N_{in}$. According to this rule, the number of outlet streams from the pinch must be greater or equal to the number of inlet streams. (In the following sections, we will explain that in case of contravention of this rule, the outlet streams must be split into different branches.)
2. We always start from the pinch point and then reach to the other side.
3. We would have pinch matches based on the number of inlet streams, which must be the same. (Pinch matches actually illustrate the exchangers that have a temperature value equal to the pinch point at their one side.)

Regarding the pinch matches, we must follow the heat capacity rule. In accordance with this law, the heat capacity of the outlet flow from the pinch point must be greater than or equal to the heat capacity of inlet flow $\left(CP_{out} \geq CP_{in} \right)$. (In the following sections we will explain that in case of contravention of this rule, the inlet streams must be split into different branches.)

Example 4.1: By using the grid diagram, design the heat exchanger's network for the process related to Example 2.1.

No of stream	Type of stream	Supply temperature (°C)	Target temperature (°C)	Heat capacity (MW/°C)
1	Hot	170	60	3
2	Hot	150	30	1.5
3	Cold	80	140	4
4	Cold	20	135	2

$$\text{Pinch point} = \begin{cases} 90\ ^\circ\text{CHot} \\ 80\ ^\circ\text{CCold} \end{cases}$$

$$Q_{H_{min}} = 20\ \text{MW}, \quad Q_{C_{min}} = 60\ \text{MW}$$

Answer: Regarding this process and in order to draw the grid diagram, we would have to draw a vertical dotted line and then write the hot temperature above it and the pinch cold temperature below it. Then we draw hot streams from left to right and the cold temperature from right to left. Moreover, we indicate heat capacities of streams at the right of the graph (see Figure 4.1).

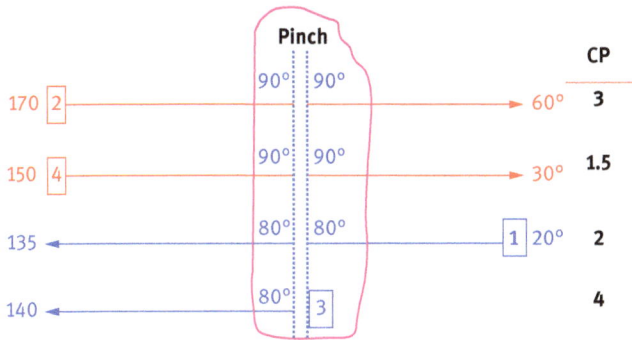

Figure 4.1: Primary grid diagram for Example 4.1.

Note that the upper and lower sides of the pinch point must be designed separately. Therefore, we shall firstly design the upper area, and all the rules of exchanger design must be performed on the upper area of pinch.

As it is clear in Figure 4.1 of upper pinch area, the inlet streams include two hot streams and the outlet streams contain two cold streams ($N_{out} = 2$, $N_{in} = 2$), so the law of streams number are all followed and applied.

1. We start from the pinch point and then we continue to the other side.
2. The number of inlet streams is 2, so two pinch matches are required.
3. According to the law of streams heat capacity, we place a pinch match between streams 2 and 3 and other pinch match between streams 4 and 1.

With reference to the explanations earlier, we would reach to Figure 4.2. (The heat load of each stream is mentioned below it.)

As shown in Figure 4.2, the amount of energy required by streams 2, 3 and 4 have been supplied, but stream 1 needs 20 MW of energy so that it receives from the hot utility.

Therefore, the design of the top area is completed (Figure 4.3). Now we start designing the heat exchangers in the lower area.

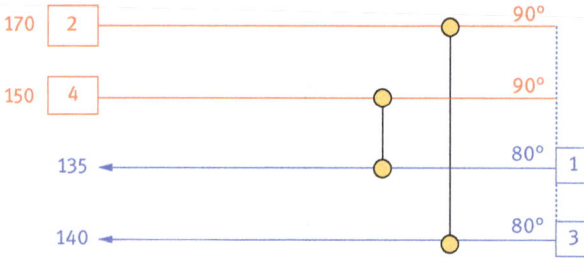

Figure 4.2: Grid diagram of pinch matches in the upper side of the pinch for Example 4.1.

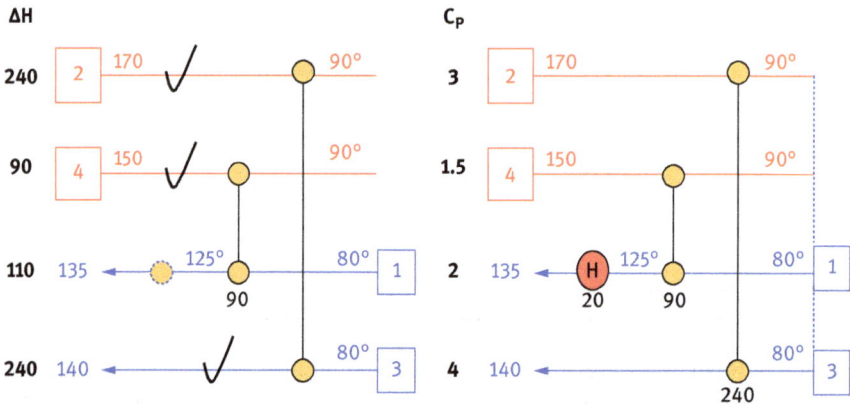

Figure 4.3: Grid diagram of the upper area regarding Example 4.1.

In order to design the lower region of the pinch, we perform every single rule again.

1. As it is obvious from the lower area of pinch in Figure 4.3, at this region the inlet stream to the pinch is only one cold stream and the number of outlet streams from the pinch is two hot streams $(N_{in} = 1, N_{out} = 2)$, so the rule for the number of streams has been applied and considered.

2. We start from the pinch point and then we continue to the other side.

3. The number of inlet stream is 1, so one pinch match is required.

4. According to the law of streams heat capacity, we can only create a pinch match between streams 1 and 2.

According to Figure 4.4, the total amount of energy required by stream 2 has been supplied but stream 1 needs 30 MW of energy and stream 4 should lose 90 MW, so we can place an exchanger between streams 1 and 4 with a heat load of 30 MW.

Therefore, the required energy for stream 1 will be supplied. (Note that it is not necessary for the other exchangers that are not considered in pinch matches to follow the rule of heat capacity.) At the end, stream 4 has to lose 60 MW of energy, and this amount could be excreted to a cold utility (Figure 4.5).

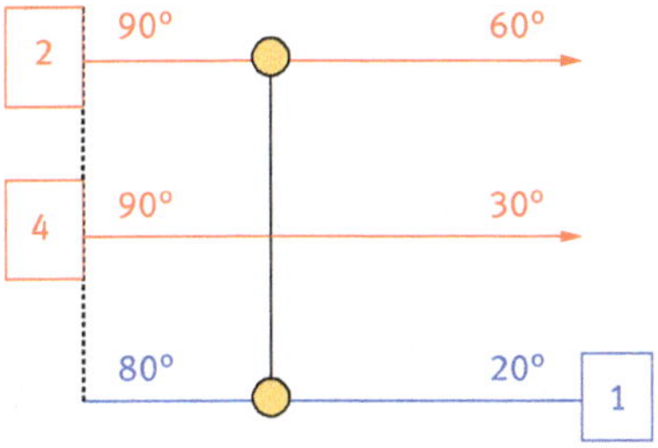

Figure 4.4: Grid diagram of pinch matches in the lower side of the pinch for Example 4.1.

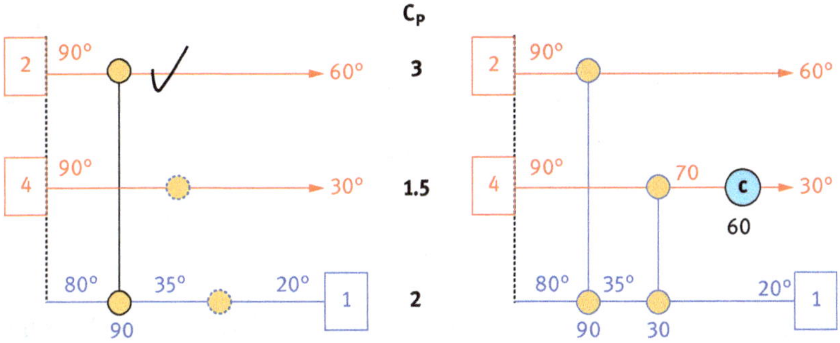

Figure 4.5: Grid diagram for the upper area of pinch regarding Example 4.1.

As a result, the design for the lower area of pinch is completed. (The heat load of the exchangers is indicated below them.)

By combining the design of both upper and lower regions of pinch point, we will achieve the design for the whole process unit. Finally, the cascade diagram of heat exchanger network could be drawn as in Figure 4.6.

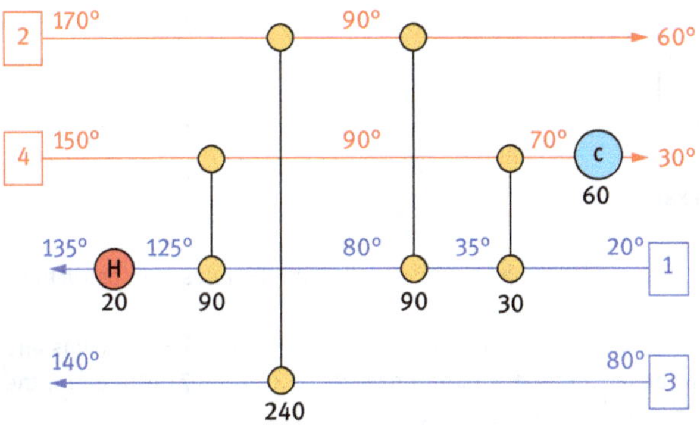

Figure 4.6: Grid diagram of Example 4.1 by MER method.

As shown in Figure 4.6, by using the pinch technology, we could apply the heat integration of this process.

Example 4.2: In accordance with the below data, design the heat exchangers network for the process belonging to Example 3.3.

No. of stream	Type of stream	Supply temperature (°C)	Target temperature (°C)	Heat capacity (MW/°C)
1	Cold	20	180	0.2
2	Cold	140	230	0.3
3	Hot	250	40	0.15
4	Hot	200	80	0.25

$$\text{Pinch point} = \begin{cases} 150\ ^\circ C\ \text{Hot} \\ 140\ ^\circ C\ \text{Cold} \end{cases}$$

$$Q_{H_{min}} = 7.5\ \text{MW}, \quad Q_{C_{min}} = 10\ \text{MW}$$

To complete the grid diagram of this unit, we draw the streams on the diagram by considering the pinch point as in Figure 4.7.

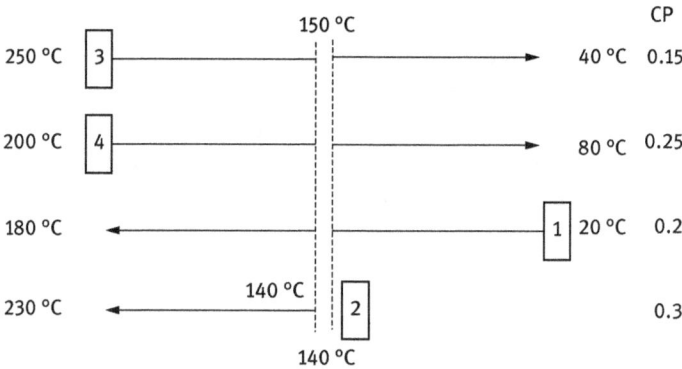

Figure 4.7: Initial grid diagram for Example 4.2.

Firstly, we start designing the upper region of pinch point, so we follow all the rules and instructions of exchanger design on this area.

1. With reference to Figure 4.7, the number of inlet flows in the upper area is two hot streams and two cold streams, and these are observed as the outlet flows ($N_{in} = 2$, $N_{out} = 2$). Therefore, the rule for the number of streams is followed.
2. We start from the pinch and then we continue to the other side.
3. The number of inlet streams is 2. So we would require two pinch matches.

4. According to the law of streams' heat capacities, there would be one pinch match between streams 2 and 4, and also other pinch match is placed between streams 1 and 3.

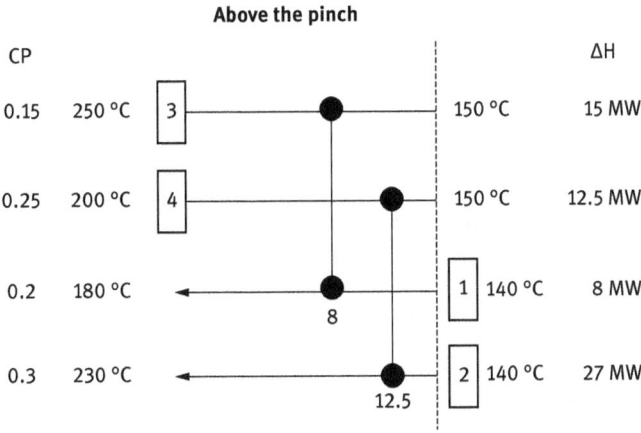

Above the pinch

CP   ΔH

0.15   250 °C   3 ——————————●—————— 150 °C   15 MW

0.25   200 °C   4 ————————————————●— 150 °C   12.5 MW

0.2   180 °C   ◄————————●————————┤ 1 │ 140 °C   8 MW
                          8

0.3   230 °C   ◄————————————————●┤ 2 │ 140 °C   27 MW
                          12.5

Figure 4.8: Grid diagram of pinch matches in the upper area of the pinch for Example 4.2.

According to the network in Figure 4.8, the required energy for streams 1 and 4 is supplied. But stream 3 requires 7 MW more energy and stream 2 needs to lose 14.5 MW of energy. So we can place an exchanger with 7 MW heating exchange between streams 2 and 3.

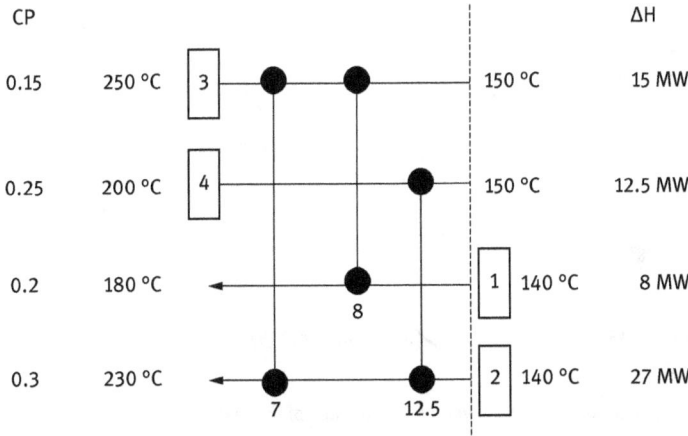

CP   ΔH

0.15   250 °C   3 ——●————————●—————— 150 °C   15 MW

0.25   200 °C   4 ————————————————●— 150 °C   12.5 MW

0.2   180 °C   ◄————————————●———————┤ 1 │ 140 °C   8 MW
                              8

0.3   230 °C   ◄——●——————————————●┤ 2 │ 140 °C   27 MW
                    7   12.5

Figure 4.9: Grid diagram of the upper area of pinch for Example 4.2 (A).

Note that for the exchangers that are not considered as pinch matches, we have to calculate the temperatures of inlet and outlet streams in order to make sure that the value of ΔT_{min} is right and not conflicted.

As shown in Figure 4.9, in the exchanger with a heating load of 7 MW, the inlet hot flow's temperature is equal to 250 °C. Meanwhile, the temperature of outlet flow from this exchanger will be calculated through the following equation, based on the heating load, heat capacity and the inlet temperature (Figure 4.10):

$$\Delta H = C_P(T_s - T_t)$$

$$7 = 0.15(250 - T_t) \rightarrow T_t = 203.33 \,°C$$

Regarding the cold flow no. 2 in this exchanger, the temperatures of inlet and outlet streams are not clear and we have no idea about them. (Note that the outlet flow temperature is not 230 °C, and this stream despite of going through an exchanger with a heating load of 7 MW would have to lose 7.5 MW more energy to reach 230 °C.) The temperature of inlet flow to the exchanger with 7 MW of heating load will be the same as the temperature of outlet flow from the exchanger with 12.5 MW of energy. Therefore, the temperature of outlet flow from the 12.5 MW exchanger is determined as follows:

$$\Delta H = C_P(T_s - T_t)$$

$$-12.5 = 0.3(140 - T_t) \rightarrow T_t = 181.66 \,°C$$

Now we can calculate the temperature of outlet flow from the exchanger with a heating load of 7 MW as follows:

$$-7 = 0.3(181.66 - T_t) \rightarrow T_t = 204.99 \,°C$$

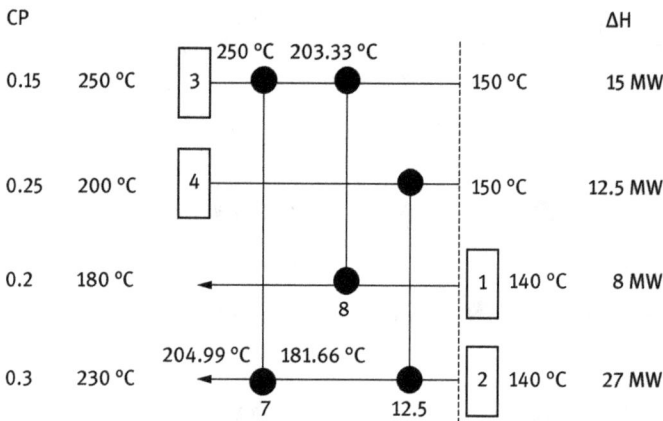

Figure 4.10: Grid diagram for the upper area of pinch for Example 4.2 (B).

Then we calculate the temperature difference between the two sides of this exchanger:

$$\begin{cases} 250 - 204.99 = 45.01 \\ 203.33 - 181.66 = 21.67 \,°C \end{cases}$$

As you see, $\Delta T_{min} = 10\,°C$ is not conflicted, and this exchanger has been located properly. In this regard, the required energy by stream 3 will be supplied but stream 2 needs 7.5 MW of energy that must be provided from the hot utility. Therefore, in general, the design of the upper side of the pinch would be as shown in Figure 4.11.

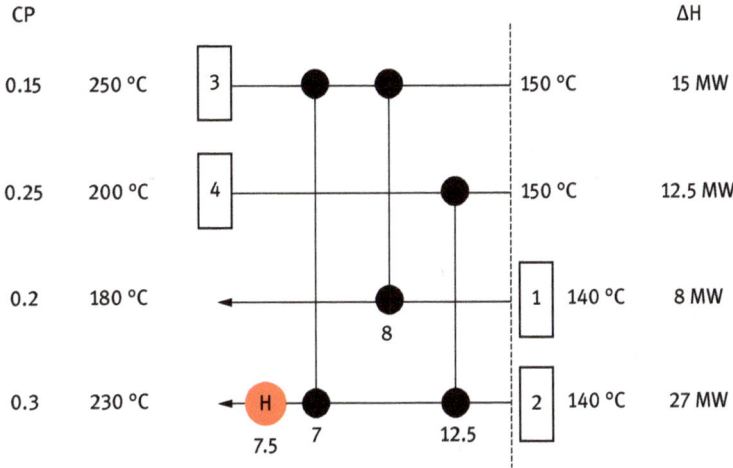

Figure 4.11: Grid diagram for the upper region of pinch regarding Example 4.2 (C).

Now we start designing the lower region of the pinch, so we apply all the instructions of exchanger design to this area.

1. As it is clear from the lower part of pinch in Figure 4.11, in this region the number of inlet streams is just one cold flow and the number of outlet flow from pinch is two hot flows ($N_{in} = 1$, $N_{out} = 2$), So the rule for number of streams is followed.
2. We start from the pinch point and then we continue to the other side.
3. The number of inlet streams to the pinch is 1, so we would require one pinch match.
4. According to the law of heat capacity of the streams, we place one pinch match between streams 4 and 1.

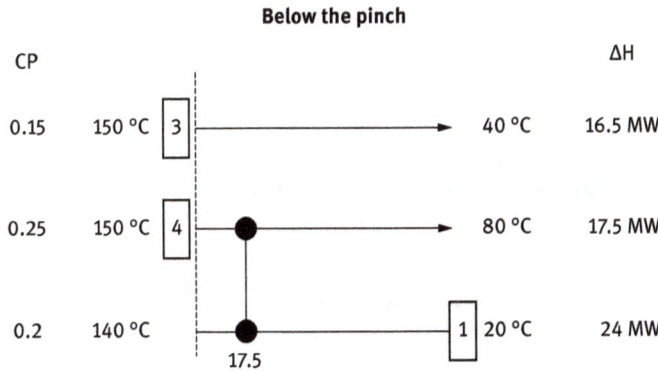

Figure 4.12: Grid diagram of pinch matches in the lower region of pinch regarding Example 4.2.

With reference to Figure 4.12, the required amount of energy by stream 4 has been supplied. But stream 1 has to lose 6.5 energy; therefore, we locate an exchanger with a heating load of 6 MW between streams 1 and 3.

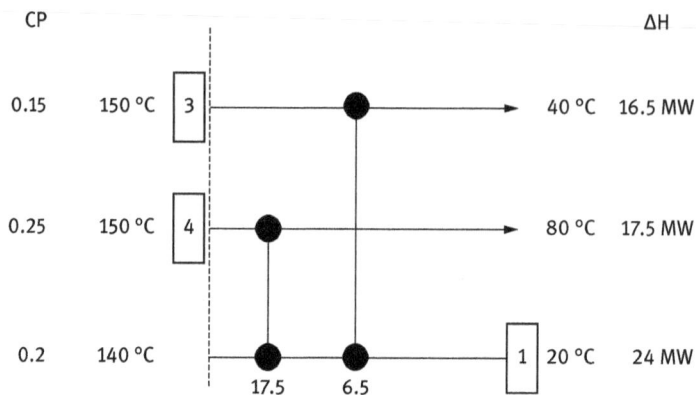

Figure 4.13: Grid diagram for the lower region of pinch regarding Example 4.2 (A).

Since this exchanger is not considered to be a pinch match, we shall calculate the temperatures around it in order to make sure that the value of ΔT_{min} is not conflicted.

As shown in Figure 4.13, the temperature of hot stream (flow no. 3) entering the exchanger with a heating load of 6.5 MW is equal to 150 °C. The outlet flow temperature from it by knowing the heating load, heat capacity and inlet flow temperature would be determined as follows:

$$\Delta H = C_P(T_s - T_t)$$
$$6.5 = 0.15(150 - T_t) \rightarrow T_t = 106.66 \ ^\circ C$$

Moreover, according to Figure 4.13, the cold flow temperature (stream no. 1) entering to this exchanger is 20 °C and the temperature value of outlet flow from the exchanger will be calculated as follows:

$$\Delta H = C_P(T_s - T_t)$$
$$-6.5 = 0.2(20 - T_t) \rightarrow T_t = 52.5 \ ^\circ C$$

The result is shown in Figure 4.14.

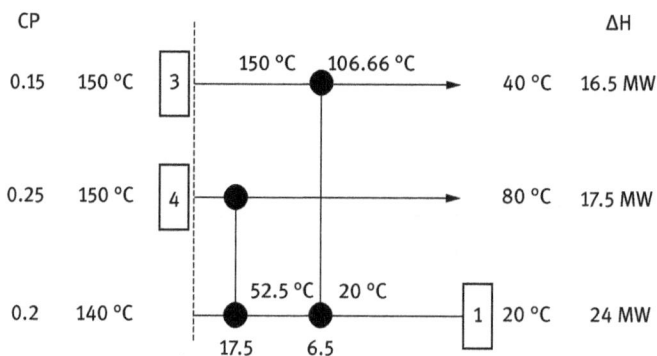

Figure 4.14: Grid diagram for the lower region of pinch regarding example.

Now we calculate the temperature difference between the two sides of an exchanger:

$$\begin{cases} 150 - 52.5 = 97.5 \ ^\circ C \\ 106.66 - 20 = 86.66 \ ^\circ C \end{cases}$$

As it turns out, the value of ΔT_{min} is not conflicted. So placing this exchanger is done correctly and stream 3 has to lose 10 MW of energy in order to reach 40 °C. Therefore, we would require a cold utility with a cooling load of 10 MW.

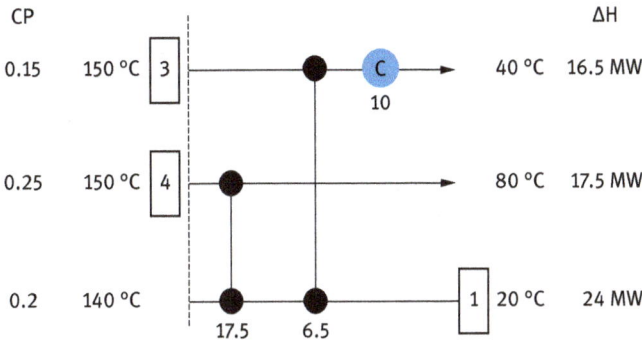

Figure 4.15: Grid diagram for the lower region of pinch regarding Example 4.2 (C).

In this regard, the design for this area is completed as shown in Figure 4.15. Now by combining the upper and lower areas of pinch, designing heat exchanger's network for the whole process will be gained as shown in Figure 4.16:

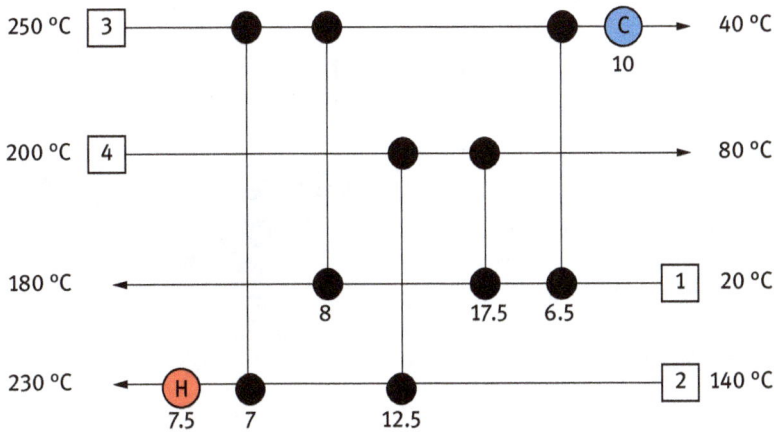

Figure 4.16: Grid diagram for Example 4.2 by MER method.

Example 4.3: Referring to the below data, design the exchanger's network for this process ($\Delta T_{min} = 50\ °C$)

Stream no.	Type of stream	Supply temperature (C)	Target temperature (C)	Heat capacity (MW/°C)
1	Hot	750	350	0.045
2	Hot	550	250	0.04
3	Cold	300	900	0.043
4	Cold	200	550	0.02

Answer: To design the heat exchanger network in this process, first of all, we must draw a cascade diagram for the unit, in order to identify the pinch points, $Q_{H_{min}}$ and $Q_{C_{min}}$.

Regarding the cascade diagram (Figures 4.17 and 4.18) and as explained earlier, while there are more than one pinch point, we have to aggregate all energies from hot utilities and enter it to the first region on top and perform heat transferring from top to the bottom of cascade again.

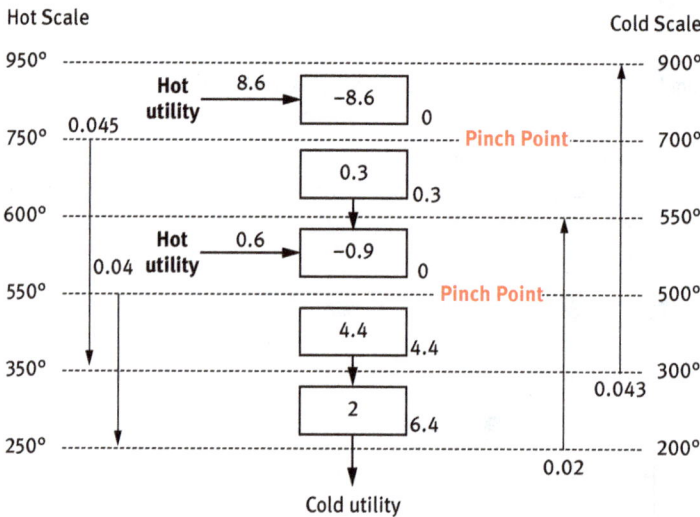

Figure 4.17: Cascade diagram for Example 4.3 (A).

So the pinch points, $Q_{H_{min}}$ and $Q_{C_{min}}$ are equal to

$$\text{Pinch point} = \begin{cases} 550\ °C\ \text{Hot} \\ 500\ °C\ \text{Cold} \end{cases}$$

$$Q_{H_{min}} = 9.2MW,\ Q_{C_{min}} = 6.4\ MW$$

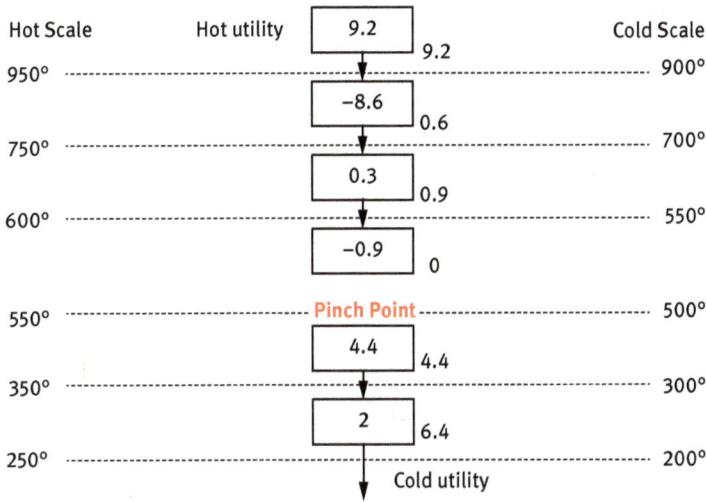

Figure 4.18: Cascade diagram for Example 4.3 (B).

Now to proceed with designing a heat exchanger network, we have to draw the grid diagram. The initial shape of the grid diagram would be as shown in Figure 4.19.

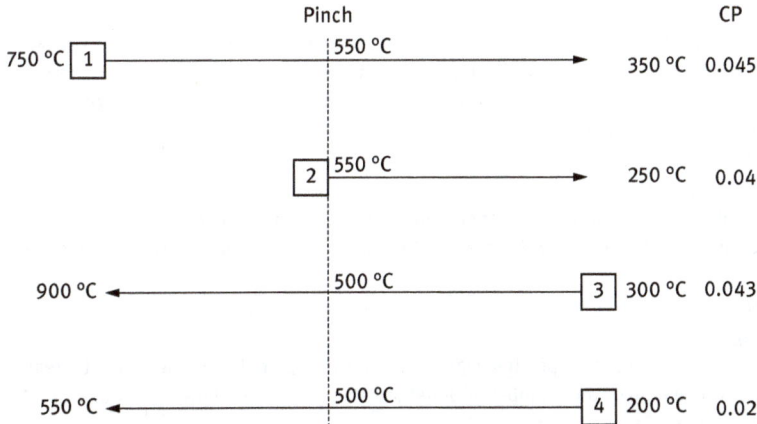

Figure 4.19: Initial grid diagram for Example 4.3.

Firstly, we start designing the upper area of pinch. In this regard, we have to perform all the rules belonging to the exchanger design at the upper region of pinch point (Figure 4.20).

1. As it is clear in Figure 4.19 at the upper area of pinch, the number of inlet streams is just the one hot stream, and the number of outlet streams from there is two cold streams ($N_{out} = 2$, $N_{in} = 1$). Therefore, the rule for the number of streams is followed and applied properly.
2. We start from the pinch side and then we continue to the other side.

3. The number of inlet flows is just 1, so we would require one pinch match.
4. In accordance to Figure 4.19, we realize that the law of heat capacity is not applied and is conflicted. So the inlet streams would have to be split.

Above the pinch

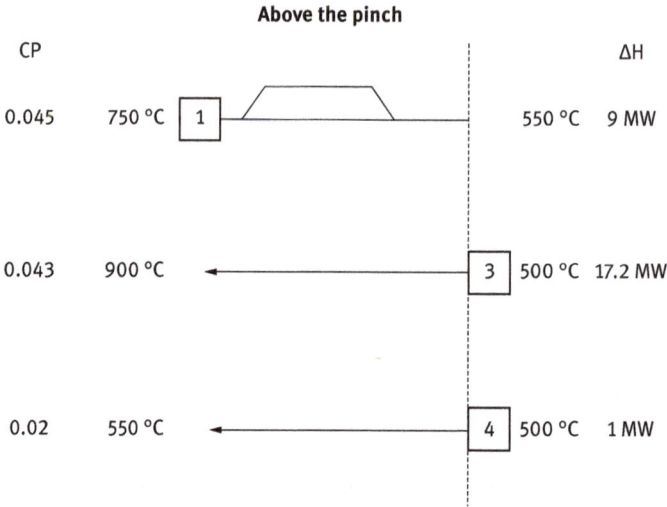

Figure 4.20: Grid diagram for the upper region of pinch regarding Example 4.3 (A).

We shall split stream 1, where the heating load of the cold streams can be fully supplied. By making this design and splitting hot stream 1 and creating an exchanger with a heating load of 1 MW between this stream and the cold stream 4, the required energy by cold stream would be completely supplied and covered. So the heat capacity of one of the branches will be 0.04 MW/°C and the other one is 0.005 MW/°C.

   This technique is named "tick off," in which we design an exchanger that by using it, a cold stream receives all its required energy or a hot stream will lose the entire energy of itself.

   Therefore, the final shape of the grid diagram for the upper region of pinch would be designed as in Figure 4.21.

   Now we proceed with designing the lower area of pinch. In this way, we apply all the instructions of exchanger design to this area.

1. As it is clear from the lower part of pinch in Figure 4.21, at this region the number of inlet streams to the pinch is two cold flows and number of outlet streams are two hot flows ($N_{in} = 2$, $N_{out} = 2$). So the rule for number of streams is followed.
2. We start from the pinch point and then we continue to the other side.
3. The number of inlet streams to the pinch is 2, so it would require two pinch matches.
4. According to the law of heat capacity for the streams, we place one pinch match between streams 1 and 3 and other pinch match between streams 2 and 4.

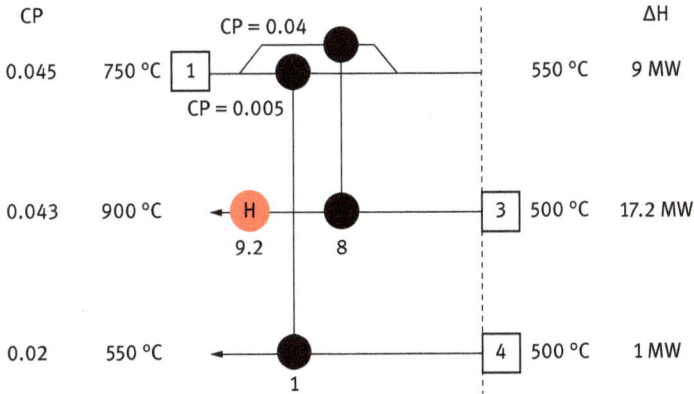

Figure 4.21: Grid diagram for the upper region of pinch regarding Example 4.3 (B).

The final design of the lower area of pinch will be as shown in Figure 4.22:

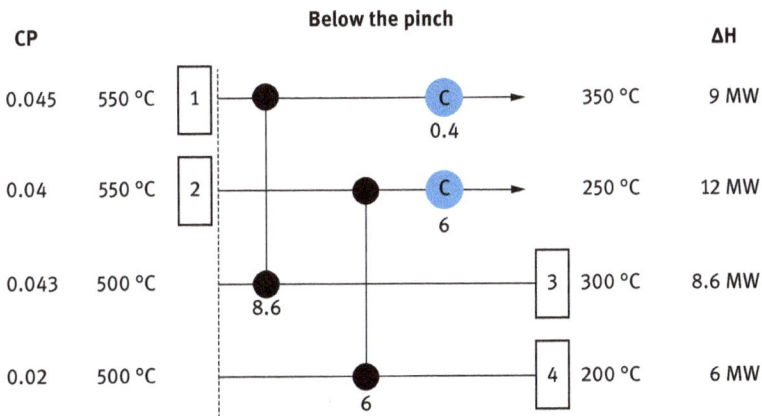

Figure 4.22: Grid diagram for the lower region of pinch regarding Example 4.3.

By combining the design of upper and lower regions of pinch, we would have the final design for the process heat exchanger network (Figure 4.23).

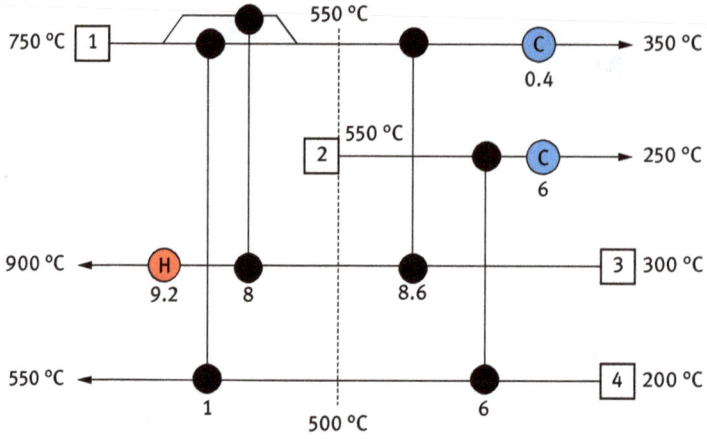

Figure 4.23: Completed grid diagram for example 4.3 with MER method.

Chapter 5
Heat Exchanger Network Design with Minimum Heat Transfer Unit

As explained earlier in the last chapter, generally, there are two methods of heat exchanger network design. We studied the method of maximum energy recovery (MER) in the previous chapter.

In this chapter, we would examine the design method of minimum process heat exchanger (N_{min}). But first of all we shall define the concept of loop in exchanger's network and the way of estimating the number of heat exchangers.

Loop definition: In case that we start moving from a point, then return to that point from the other route so that we would have a loop in the network.

5.1 Heat Exchanger Network Design with Minimum Heat Transfer Unit (N_{min})

This method is showing the network with minimum number of equipment and heat exchangers that in this case the energy load of utilities will increase.

In order to reduce the number of heat exchangers, we have to identify the loops in the grid diagram and remove them. By removing loops, the first golden rule of pinch will be conflicted and some amount of energy will pass through the pinch point, which may cause an increase in energy load of hot and cold utilities. Therefore, the number of heat exchangers could be determined by two exchanger design methods as follows:

Number of exchangers by MER design = Number of streams + Number of heat sources + Number of loops − Independent systems

Number of heat exchangers by N_{min} design = Number of streams + Number of heat sources − Independent systems

Therefore, to design a heat network by minimum number of heat exchangers, the capital cost decreases and it will cause an increase in energy cost. Actually, a network might comprise several loops, but to answer this question that how many loops shall be removed would depend on the interactions between capital cost and energy cost and by considering the general aims of the project we need to investigate this issue.

Heat exchanger network design by minimum number of heat exchanger (N_{min}) has lower capital costs and more energy costs in comparison to the network design by MER method. Therefore, a heat exchanger network design through the method of minimum number of heat exchangers (N_{min}) would be a suitable choice for the projects that are facing limitations and lack of initial capital cost but the energy cost is in a minimum level in that country.

https://doi.org/10.1515/9783110786323-005

Example 5.1: Design the heat exchanger network by N_{min} method for the following process unit (Example 1.2).

Stream no.	Type of stream	Supply temperature (°C)	Target temperature (°C)	Heat capacity (MW/°C)
1	Hot	170	60	3
2	Hot	150	30	1.5
3	Cold	80	140	4
4	Cold	20	135	2

$$\text{Pinch point} = \begin{cases} 90\,°C & \text{Hot} \\ 80\,°C & \text{Cold} \end{cases}$$

$$Q_{H_{min}} = 20\ \text{MW},\ Q_{C_{min}} = 60\ \text{MW}$$

Answer: As you can see, the grid diagram of this unit in Example 4.1 was shown based on the aim of MER as in Figure 5.1.

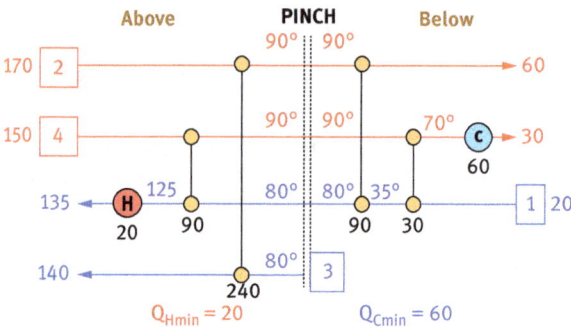

Figure 5.1: Grid diagram for Example 5.1 by MER method.

First of all, we shall identify the loop and then mark it by a dotted line (Figure 5.2).

Generally, between the exchangers involving in a loop, we remove the one which has the minimum heating load. So we remove the exchanger with 30 MW of energy and we sum up this amount with the exchanger with 90 MW of energy.

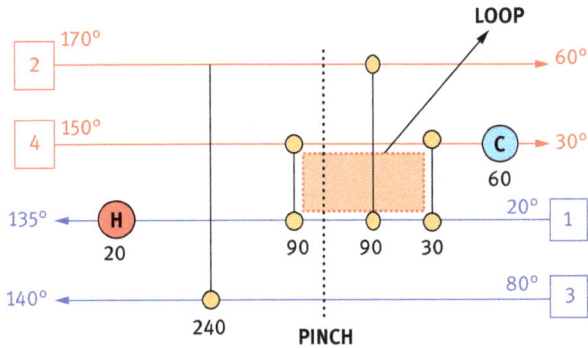

Figure 5.2: Identifying the loop in a grid diagram for Example 5.1.

Now we calculate the inlet and outlet temperatures from the 120 MW exchanger, which is shown in Figure 5.3.

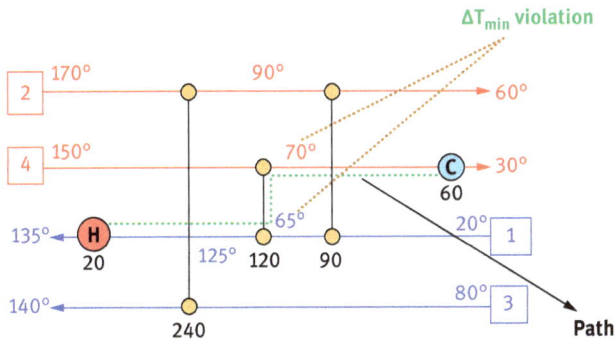

Figure 5.3: Breaking the loop in a grid diagram and making path in Example 5.1 (A).

As shown in Figure 5.3, the temperature difference at one side of the exchanger is equal to 5 °C, which means that the value of ΔT_{min} has been conflicted. In order to modify this model, we need to reduce the heating load of 120 MW to make the temperature difference at the right of the exchanger reach 10 °C. In this way, the first law of pinch will conflict and as much as reducing the heat load of the exchanger should be added to the energy load of cold and hot utilities. In this regard, we define path 1.

A path is a connection which is made between two hot and cold utilities so that it passes through the defective exchanger and resolves the problems. (Usually after breaking loops in the heat exchanger network, some heat paths will be made in that network.)

This issue is shown in Figure 5.4.

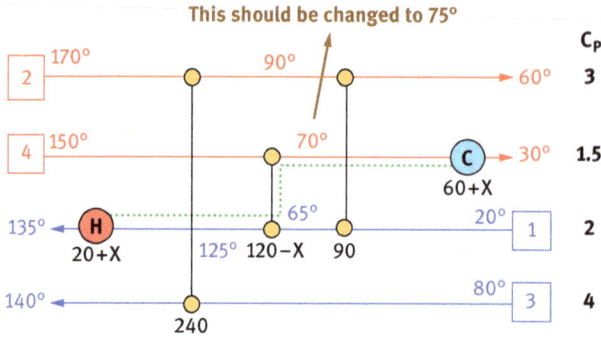

Figure 5.4: Breaking the loop in a grid diagram and making path in Example 5.1 (B).

X will be calculated through the following equation:

$$120 - X = 1.5 (150 - 75) \rightarrow X = 7.5$$

So by minimum number of heat exchangers (N_{min}) method, the design will be completed (Figure 5.5).

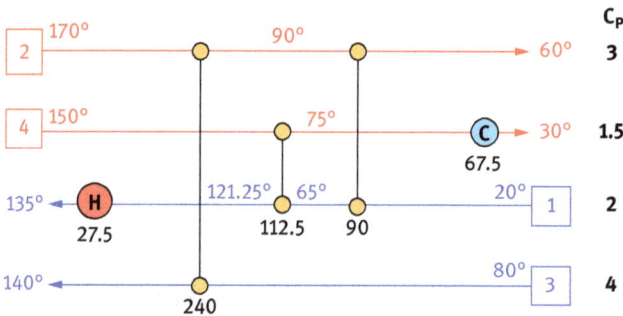

Figure 5.5: Grid diagram of Example 5.1 by minimum number of heat exchanger (N_{min}) method.

5.2 Grand Composite Curve (GCC)

Utilities consist of different types, as follows:

Hot utilities: 1 – furnace; 2 – low-pressure steam, medium-pressure steam, high-pressure steam (HP); 3 – hot oil utility.

Cold utilities: 1 – cooling water; 2 – cooling air; 3 – refrigeration system.

The point is that the most economical and suitable utilities must be selected for heat exchanger network design.

The cascade and grid diagrams will show the pinch point and the values of $Q_{H_{min}}$ and $Q_{C_{min}}$, but they cannot identify the types of hot and cold utilities. In order to choose the suitable type of utilities we shall use the grand composite curve (GCC).

To draw the GCC, we would follow the instructions as follows:

1. Firstly, we draw the cascade diagram.
2. In order to calculate the interval temperature (T^*) for each region, we would use one of the following equations:

$$\begin{cases} T^* = T_{Cold} + \frac{\Delta T_{min}}{2} \\ T^* = T_{hot} - \frac{\Delta T_{min}}{2} \end{cases}$$

3. We extract the enthalpy of each region from the cascade diagram and then calculate the enthalpy value corresponding to each interval temperature. In this purpose, the enthalpy of the first area above the pinch in the cascade diagram is $Q_{H_{min}}$, and the value for first area below the pinch point is zero.
4. In GCC, the vertical axis is the interval temperature (T^*) and the horizontal axis is the enthalpy (H). Therefore, by having the interval temperature and enthalpy of each region, a point on the graph will be specified.

This must be applied to all regions and all the points must be connected by a line.

So in this regard, GCC is drawn and the shape would be as shown in Figure 5.6.

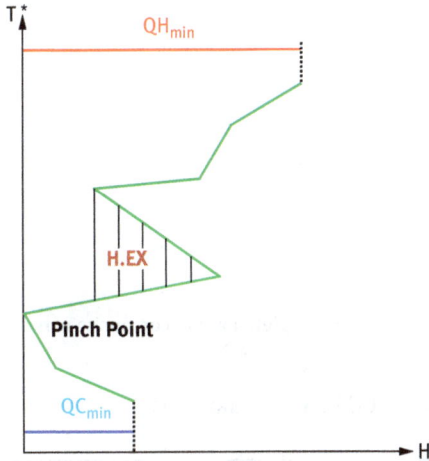

Figure 5.6: Initial grand composite curve (GCC).

The GCC expresses the following items:

1. The intersection of the curve with the vertical axis indicates the pinch point.
2. The GCC would divide the whole process into upper and lower sides of the pinch as well as the composite curve.
3. The top open portion of graph shows the value of $Q_{H_{min}}$.

4. The bottom portion of graph will indicate the value of $Q_{C_{min}}$.
5. The positive slope lines indicate cold streams and negative slope lines show hot streams.
6. Hatching areas represent the amount of heat transfer between processes. In fact, in these areas, negative slope lines provide the required heat for positive slope lines.

Now the question is that, how could we select the most suitable utility by the help of GCC diagram?

To answer this question, firstly note Figure 5.7.

As it is obvious in the diagram, heating load of the hot utility is equal to $Q_{H_{min}}$ that this amount would be supplied by HP steam.

With the help of GCC, we can use steam at three different levels; instead of using HP steam, this will cause reduction in energy costs.

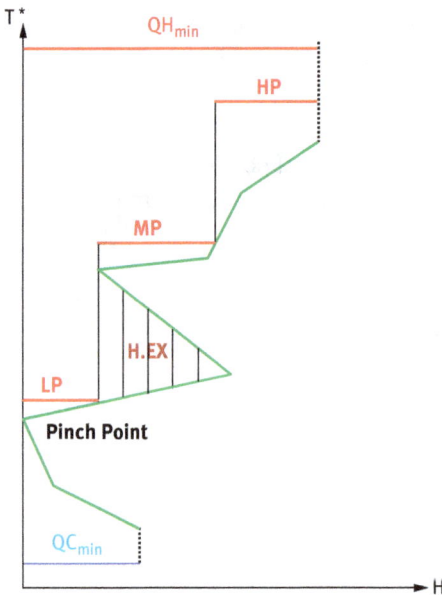

Figure 5.7: Completed grand composite curve.

In the following, we will explain how to draw a GCC in a few examples.

Example 5.2: Draw the GCC for Example 5.1.
Answer: Firstly, we have to draw the cascade diagram (Figure 5.8), which we have already done for this process unit in Chapter 3 (Example 2.3).

At the first region in the upper side of the pinch $T_{cold} = 160\ °C$, $T_{hot} = 170\ °C$ and $Q = 20$ MW, so the interval temperature for this area is equal to:

$$T^* = 170 - \frac{10}{2} = 165\ °C \text{ or } T^* = 160 + \frac{10}{2} = 165\ °C$$

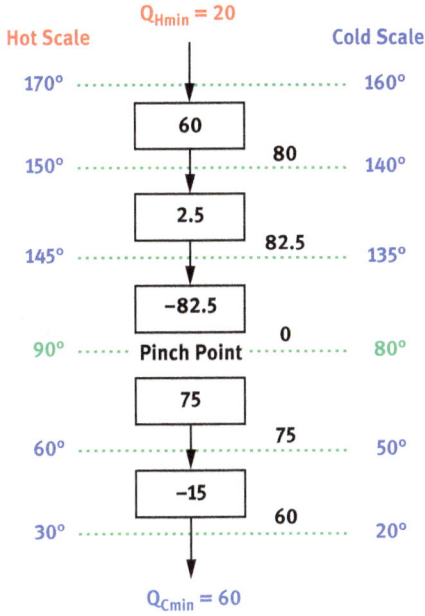

Figure 5.8: Cascade diagram for Example 5.2.

In the same way, we calculate the interval temperature (T^*) of each area created in the cascade diagram, and results are shown in the following table:

Region	Interval temperature (°C)	Enthalpy (MW)
1	165	20
2	145	80
3	140	82.5
4	85	0
5	55	75
6	25	60

With accordance to this table, we can draw GCC.

According to Figure 5.9, the values of $Q_{C_{min}}$, $Q_{H_{min}}$ and pinch point are equal to

$$\text{Pinch point} = \begin{cases} T_{hot} = 85 + \frac{10}{2} = 90\,°C \\ T_{cold} = 85 - \frac{10}{2} = 80\,°C \end{cases}$$

$$Q_{H_{min}} = 20\text{ MW}, \quad Q_{C_{min}} = 60\text{ MW}$$

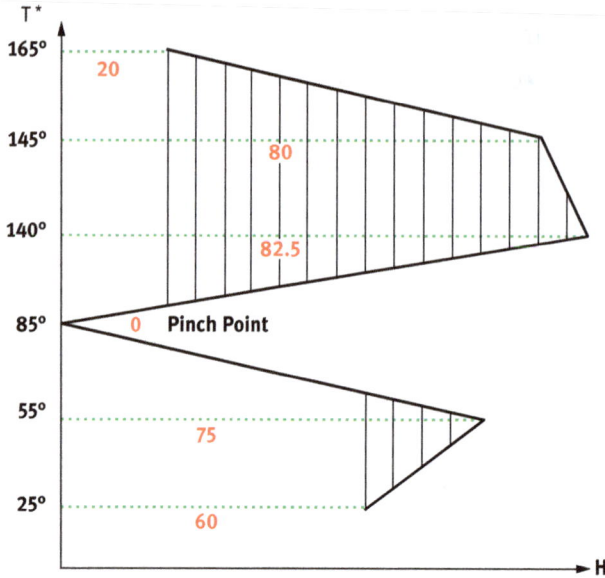

Figure 5.9: Grand composite curve for Example 5.2.

Example 5.3: Draw GCC for the below process unit (Example 3.3) ($\Delta T_{min} = 10\,°C$).

Stream no.	Type of stream	Supply temperature (°C)	Target temperature (°C)	Heat capacity (MW/°C)
1	Cold	20	180	0.2
2	Cold	140	230	0.3
3	Hot	250	40	0.15
4	Hot	200	80	0.25

Answer: First of all, we have to draw the cascade diagram (Figure 5.10), which already has been drawn for this process unit in Chapter 3.

At the first area, $T_{cold} = 240\,°C$, $T_{hot} = 250\,°C$, $Q = 7.5\,MW$; so the interval temperature in this area will be calculated through the following equation:

$$T^* = 240 - \frac{10}{2} = 245\,°C \text{ or } T^* = 240 + \frac{10}{2} = 245\,°C$$

In the same exact way, we determine the interval temperatures (T^*) of the other regions, and the results are shown in the following table:

Hot Scale   Hot utility   7.5   Cold Scale

250°		7.5	7.5	240°
240°		1.5	9	230°
200°		−6	3	190°
190°		1	4	180°
150°		−4	0	140°
		Pinch Point		
80°		14	14	70°
40°		−2	12	35°
30°		−2	10	20°

▼ Cold utility

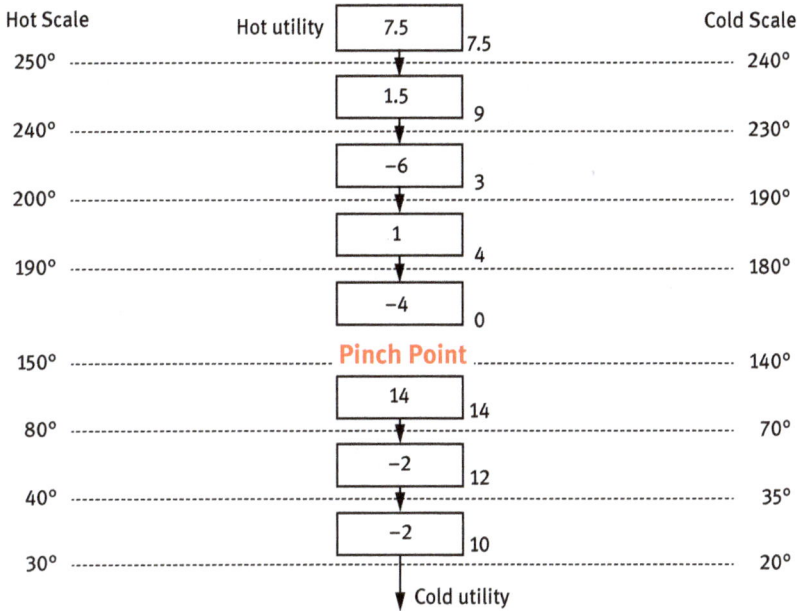

Figure 5.10: Cascade diagram for Example 5.3.

Region	Interval temperature (°C)	Enthalpy (MW)
1	245	7.5
2	235	9
3	195	3
4	185	4
5	145	0
6	75	14
7	35	12
8	25	10

With reference to this table, we draw the GCC as in Figure 5.11. The values of $Q_{C_{min}}$, $Q_{H_{min}}$ and pinch point are equal to

$$\text{Pinch point} = \begin{cases} T_{hot} = 145 + \frac{10}{2} = 150\,^\circ C \\ T_{cold} = 145 - \frac{10}{2} = 140\,^\circ C \end{cases}$$

$$Q_{H_{min}} = 7.5\ \text{MW},\ Q_{C_{min}} = 10\ \text{MW}$$

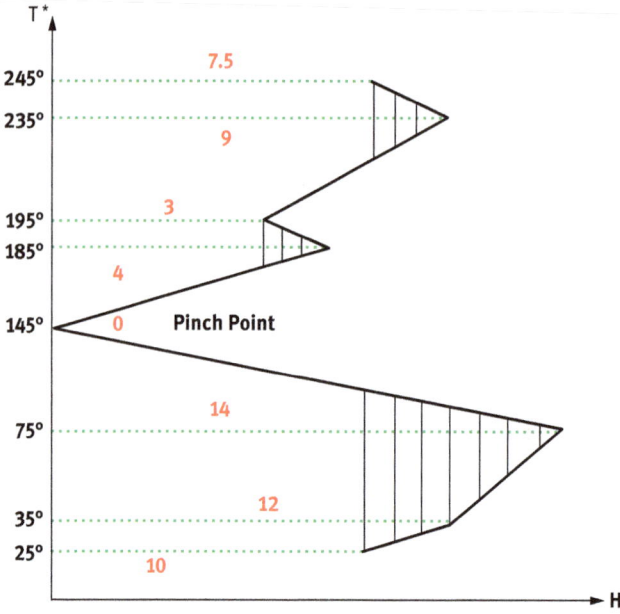

Figure 5.11: Grand composite curve for Example 5.3.

Chapter 6
Estimating the Number of Shells and Heat Transfer Area of Heat Exchanger Network

6.1 Area Targeting

In addition to the required information to predict energy targeting, the composite curve also contains some necessary data for area targeting of the heat exchanger network [11].

To calculate the surface area of heat exchanger network by using a composite curve, we must also consider the streams of utilities as well as process streams for plotting the curves, and for drawing the composite curve, we would use the cumulative enthalpy method.

In order to determine the cumulative enthalpy of hot streams, we consider the lowest temperature of the hot stream as the basis and set its cumulative enthalpy to zero. And to calculate the cumulative enthalpy of other temperature ranges in hot streams, we evaluate the enthalpy of each temperature range and then add the enthalpy of the previous range to obtain the cumulative enthalpy of that range [12].

In order to determine the cumulative enthalpy of cold streams, we consider the lowest temperature of the cold stream as the basis and set its cumulative enthalpy to the heating load of cold utility. And to calculate the cumulative enthalpy of other temperature ranges in cold streams, we evaluate the enthalpy of other temperature range and then add the enthalpy of the previous range to obtain the cumulative enthalpy of that range [13].

Example 6.1: Draw the composite curve for the process unit data in the following table (Example 2.1) by using the cumulative enthalpy method.

Stream no.	Type of stream	Supply temperature (°C)	Target temperature (°C)	Heat capacity (MW/°C)
1	Hot	170	60	3
2	Hot	150	30	1.5
3	Cold	80	140	4
4	Cold	20	135	2

The values of $Q_{C_{min}}$, $Q_{H_{min}}$ and pinch point are equal to

https://doi.org/10.1515/9783110786323-006

$$\text{Pinch point} = \begin{cases} 90\,^\circ\text{C Hot} \\ 80\,^\circ\text{C Cold} \end{cases}$$

$$Q_{H_{min}} = 20 \text{ MW}, \ Q_{C_{min}} = 60 \text{ MW}$$

Answer: Firstly, draw the cascade diagram (Figure 6.1) for this unit and then calculate the cumulative enthalpy of the cold and hot utilities separately.

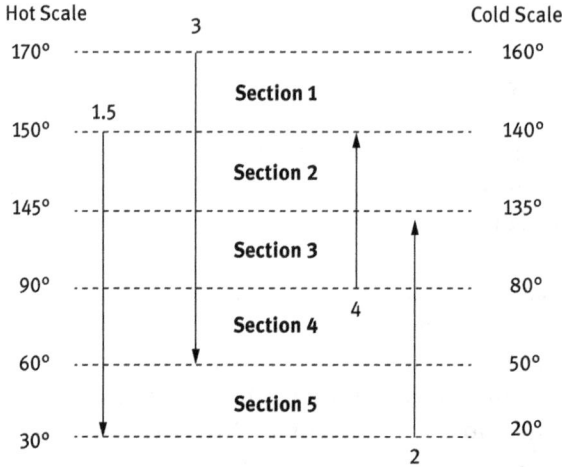

Figure 6.1: Cascade diagram of streams for Example 6.1.

The cumulative enthalpy of hot streams will be calculated through the following instruction:

Temperature (°C)	Enthalpy (MW)	Cumulative enthalpy (MW)
30	0	0
60	1.5 (60 − 30) = 45	0 + 45 = 45
150	(1.5 + 3) (150 − 60) = 405	405 + 45 = 450
170	3 (170 − 150) = 60	60 + 450 = 510

The cumulative enthalpy of cold streams will be calculated as follows:

Temperature (°C)	Enthalpy (MW)	Cumulative enthalpy (MW)
20	60	60
80	2 (80 − 20) = 120	60 + 120 = 180
135	(4 + 2) (135 − 80) = 330	180 + 330 = 510
140	4 (140 − 135) = 20	510 + 20 = 530

As a result, the composite curve (cumulative enthalpy) for this unit will be plotted in Figure 6.2.

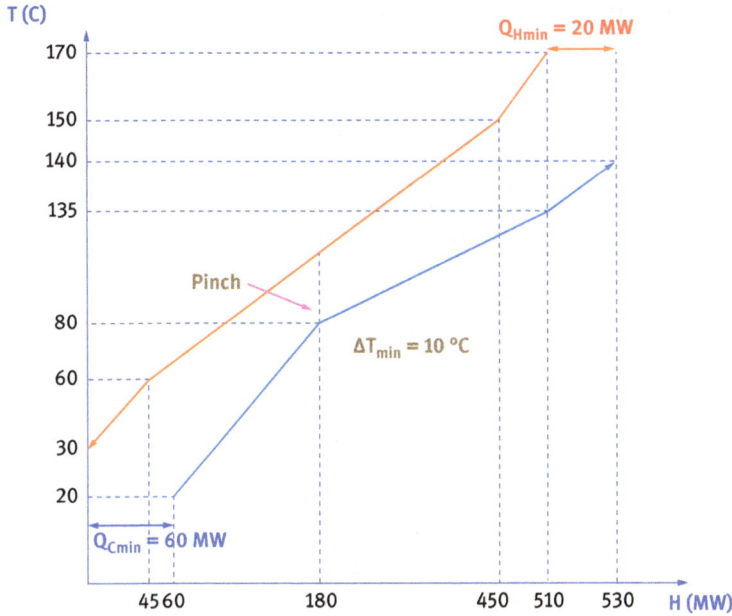

Figure 6.2: Composite curve (cumulative enthalpy) for Example 6.1.

Regarding area targeting, we shall draw the composite graphs, then we place a vertical line at every fracture. In this way, the curves and the surface area between two graphs of hot and cold streams are divided into parts and intervals consisting of enthalpies. Then we calculate the minimum of surface area between each interval through the following equation:

$$A_k = \frac{1}{(\text{LMTD})_k (F_T)_k} \left(\sum_{i=0}^{n} \frac{Q_i}{h_i} \right)$$

Hence, before completing the design of whole network, by assuming that all heat exchangers in the network are made of the same material, the minimum surface area of exchangers is equal to

$$A_{min} = \sum A_k$$

where Q_i is the heat load of each stream, h_i is the heat transfer coefficient of each stream, n is the number of streams in every interval (the streams in every interval can be included of hot streams, cold streams as well as hot and cold utilities), F_T is the correction factor in every interval, LMTD is the log mean temperature difference in every interval and K is the counter in every interval.

6.2 Log Mean Temperature Difference (LMTD) Calculation

Logarithmic mean temperature difference between hot and cold streams will be determined as follows:

Cocurrent

$$\Delta T_{\text{LMTD}} = \frac{(T_{h\,in} - T_{c\,in}) - (T_{h\,out} - T_{c\,out})}{\ln \dfrac{T_{h\,in} - T_{c\,in}}{T_{h\,out} - T_{c\,out}}}$$

$$T_{h\,in} \rightarrow T_{h\,out}$$

$$T_{c\,in} \rightarrow T_{c\,out}$$

Countercurrent

$$\Delta T_{\text{LMTD}} = \frac{(T_{h\,in} - T_{c\,out}) - (T_{h\,out} - T_{c\,in})}{\ln \dfrac{T_{h\,in} - T_{c\,out}}{T_{h\,out} - T_{c\,in}}}$$

$$T_{h\,in} \rightarrow T_{h\,out}$$

$$T_{c\,in} \leftarrow T_{c\,out}$$

6.3 Estimating Correction Factor (FT)

Shell and tube heat exchangers are being used in chemical process industrial applications more than the other types. The simplest type of them is 1–1 shell and tube exchanger (which means one shell pass and one tube pass). Shell pass and tube pass are presenting the number of passes which are supposed to be in a shell and in a tube. Figure 6.3 shows this kind of heat exchanger as well as the temperature difference diagram in comparison to the length of exchanger as in Figure 6.4.

The fluid stream in a shell and tube exchanger 1–1 is countercurrent. Because of this, the correction factor would be one and the equation of heating load for this exchanger will be as follows:

$$Q = UA\Delta T$$

where Q is the heating load of exchanger, U is the heat transfer coefficient of the exchanger, ΔT_{LMTD} is the logarithmic temperature difference and A is the exchanger surface area.

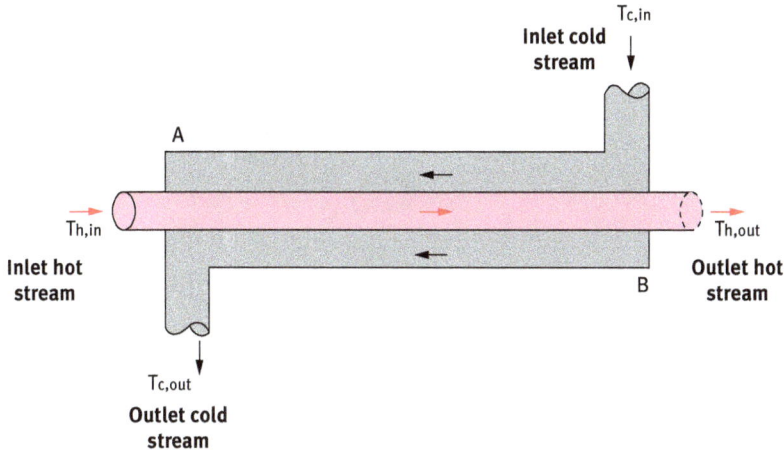

Figure 6.3: 1–1 Shell and tube exchanger (one pass shell and one pass tube).

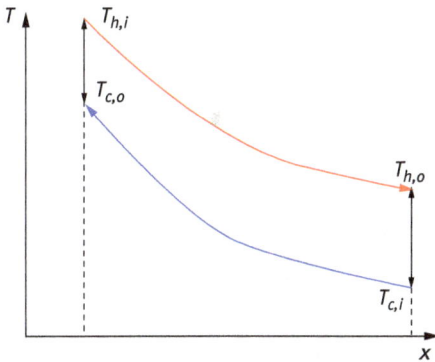

Figure 6.4: Diagram of temperature difference – length of shell and tube heat exchanger 1–1.

In spite of having minimum surface area of heat transferring in shell and tube heat exchanger 1–1, with a certain heating load and heat transfer coefficient, but they have some major weaknesses.
1. The tube's bundle is fixed and cannot be removed from the shell for physical or chemical washing.
2. With the fixed tube bundle, a slight expansion of the shell and tube will cause damage to the exchanger.
3. In this type of exchanger, there should be a temperature cross.

There are some other arrangements for shell and tube exchangers, and the most common and widely used type is 1–2 exchanger (one pass shell and two passes of tubes). Figure 6.5 shows this kind of exchanger and the diagram of temperature difference–pipe length in the following section.

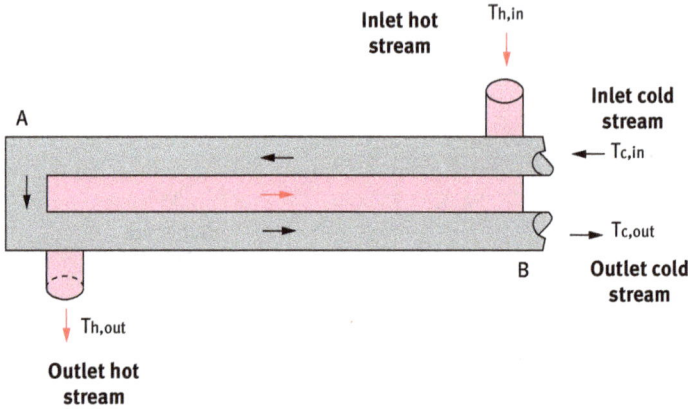

Figure 6.5: 1–2 Shell and tube exchanger (one pass shell and two passes of tubes).

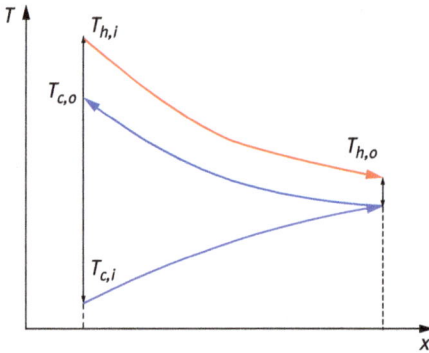

Figure 6.6: Diagram of temperature difference versus shell and tube exchanger length.

As shown in Figure 6.6, in 2–1 shell and tube exchanger, some of the streams are countercurrent and the others are cocurrent. Due to this matter, the effects of temperature difference on heat transferring will reduce in comparison to the complete countercurrent streams. To take this into consideration for network design, we use a numerical correction factor between zero and one. Therefore, the heat load relation for these exchanger is as follows:

$$Q = F_T U A \Delta T$$

where Q is the heat load, U is the heat transfer coefficient of exchanger, ΔT_{LMTD} is the log mean temperature difference, A is the exchanger surface area and F_T is the correction factor.

2–1 Shell and tube exchangers have so many industrial advantages that have made them become very important and widely used in industries. Some of these benefits are:

1. Easy physical or chemical washing of these exchangers
2. Thermal expansion capability
3. Having appropriate heat transfer coefficient of fluid inside the tube due to its high speed

Generally, the correction factor (F_T) depends on two thermal elements P and R. The relation between P and R is as follows:

$$R = \frac{T_{h\,in} - T_{h\,out}}{T_{c\,out} - T_{c\,in}}$$

$$P = \frac{T_{c\,out} - T_{c\,in}}{T_{h\,in} - T_{c\,in}}$$

By identifying the inlet and outlet temperatures, we can determine P and R parameters, and then we would be able to calculate the value of correction factor by using the graphs in heat transfer references.

Example 6.2: By assuming that all exchangers have been supplied with the same material, determine the minimum required heat transferring surface area of the network for the following process unit (Example 2.1). Correction factor in each interval (F_T) is equal to 1, and the other data regarding convective heat transfer coefficient of the streams and data belonging to hot and cold utilities are shown in the following table:

Stream number	Type of stream	Supply temperature (°C)	Target temperature (°C)	Heat capacity (MW/°C)	Conductive heat transfer coefficient (MW/m² °C)
1	Hot	170	60	3	0.0045
2	Hot	150	30	1.5	0.005
3	Cold	80	140	4	0.004
4	Cold	20	135	2	0.0035

We would use cooling water as cold utility by considering the inlet temperature as 20 °C, outlet temperature as 30 °C and heat transfer coefficient as 0.00375 MW/m² °C. Also for hot utility we will use steam with inlet and outlet temperatures for 250 °C and heat transfer coefficient as 0.006 MW/m² °C.

Answer: To solve this exercise we assume 1–1 shell and tube exchanger.

First Case: Before Design

As mentioned earlier in this case, the exchanger surface area would be determined from the composite curve. We once plotted a composite curve regarding the network of this process unit in Example 1.6 by using the cumulative enthalpy method.

In order to estimate the minimum surface area of heat network before designing it, we shall draw vertical lines at the point of fractures on hot and cold stream graphs to divide the diagram into thermal intervals.

The area between hot and cold graphs in the composite curve (cumulative enthalpy) is divided into six intervals (Figure 6.7).

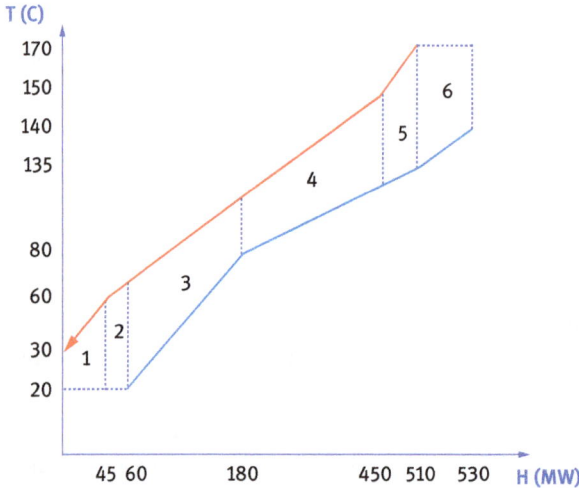

Figure 6.7: Intervals in composite curve (cumulative enthalpy) for Example 6.2.

Now we calculate the area of each interval by using the composite curve and at the end, we sum up all surfaces in order to determine the minimum required heat transfer surface area before designing the network:

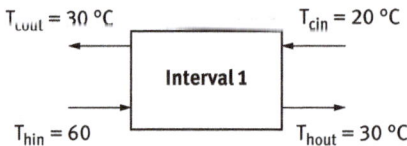

$$LMTD = \frac{(60-30)-(30-20)}{\ln\dfrac{60-30}{30-20}} = 18.2$$

$$Q_h = 1.5 \ (60-30) = 45, \quad A_1 = \left(\frac{45}{5\times 10^{-3}} + \frac{45}{3.75\times 10^{-3}}\right)\frac{1}{18.2} = 1,153.84 \ m^2$$

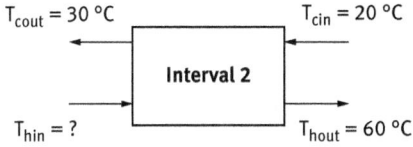

$T_{cout} = 30\ °C$　　　　　$T_{cin} = 20\ °C$

Interval 2

$T_{hin} = ?$　　　　　$T_{hout} = 60\ °C$

$$Q = FC_P\Delta T_h$$

$$(60 - 45) = (1.5 + 3)\Delta T_h \rightarrow \Delta T_h = 3.33$$

$$Q_{1h} = 1.5 \times 3.33 = 5,\ \ Q_{2h} = 3 \times 3.33 = 9.99 \approx 10$$

$$\Delta T_h = 3.33 = (T_{hin} - 60) \rightarrow T_{hin} = 63.33\ °C$$

$$LMTD = \frac{(63.33 - 30) - (60 - 20)}{\ln\dfrac{63.33 - 30}{60 - 20}} = 36.56$$

$$A_2 = \left(\frac{5}{5 \times 10^{-3}} + \frac{10}{4.5 \times 10^{-3}} + \frac{15}{3.75 \times 10^{-3}}\right) \times \frac{1}{36.56} = 197.54\ m^2$$

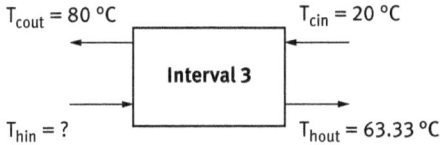

$T_{cout} = 80\ °C$　　　　　$T_{cin} = 20\ °C$

Interval 3

$T_{hin} = ?$　　　　　$T_{hout} = 63.33\ °C$

$$Q = FC_P\Delta T_h$$

$$(180 - 160) = (1.5 + 3)\Delta T_h \rightarrow \Delta T_h = 26.67$$

$$Q_{1h} = 3 \times 26.67 = 80,\ \ Q_{2h} = 1.5 \times 26.67 = 40$$

$$\Delta T_h = 26.67 = (T_{hin} - 63.33) \rightarrow T_{hin} = 90\ °C$$

$$LMTD = \frac{(63.33 - 20) - (90 - 80)}{\ln\dfrac{63.33 - 20}{90 - 80}} = 22.73$$

$$A_3 = \left(\frac{80}{4.5 \times 10^{-3}} + \frac{40}{5 \times 10^{-3}} + \frac{120}{3.5 \times 10^{-3}}\right) \times \frac{1}{22.73} = 2,642.47\ m^2$$

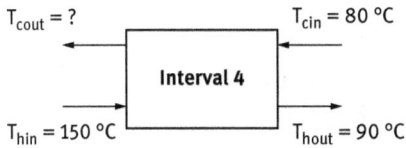

$T_{cout} = ?$　　　　　$T_{cin} = 80\ °C$

Interval 4

$T_{hin} = 150\ °C$　　　　　$T_{hout} = 90\ °C$

$$Q = FC_P\Delta T_h$$

$$(450 - 180) = (2 + 4)\Delta T_c \rightarrow \Delta T_c = 45$$

$$Q_{1c} = 2 \times 45 = 90,\ \ Q_{2c} = 4 \times 45 = 180$$

$$Q_{1h} = 1.5 \times 60 = 90,\ \ Q_{2h} = 3 \times 60 = 180$$

$$\Delta T_c = 45 = (T_{cout} - 80) \rightarrow T_{cout} = 125\ °C$$

$$LMTD = \frac{(150-125)-(90-80)}{\ln\dfrac{150-125}{90-80}} = 16.37$$

$$A_4 = \left(\frac{90}{3.5\times10^{-3}} + \frac{180}{4\times10^{-3}} + \frac{90}{5\times10^{-3}} + \frac{180}{4.5\times10^{-3}}\right) \times \frac{1}{16.37} = 7,862.81 \ m^2$$

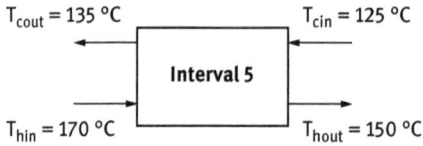

$T_{cout} = 135\ °C \qquad T_{cin} = 125\ °C$

Interval 5

$T_{hin} = 170\ °C \qquad T_{hout} = 150\ °C$

$$\Delta T_c = 10 \rightarrow Q_{1c} = 2\times10 = 20, \quad Q_{2c} = 4\times10 = 40, \quad Q_h = 3\times(150-170) = 60$$

$$LMTD = \frac{(170-135)-(150-125)}{\ln\dfrac{170-135}{150-125}} = 29.72$$

$$A_5 = \left(\frac{20}{3.5\times10^{-3}} + \frac{40}{4\times10^{-3}} + \frac{60}{4.5\times10^{-3}}\right) \times \frac{1}{29.72} = 977.37 \ m^2$$

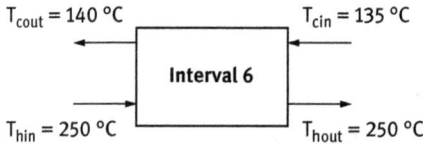

$T_{cout} = 140\ °C \qquad T_{cin} = 135\ °C$

Interval 6

$T_{hin} = 250\ °C \qquad T_{hout} = 250\ °C$

$$\Delta T_c = 5 \rightarrow Q_c = 5\times4 = 20$$

$$LMTD = \frac{(250-135)-(250-140)}{\ln\dfrac{250-135}{250-140}} = 112.48$$

$$A_6 = \left(\frac{20}{6\times10^{-3}} + \frac{20}{4\times10^{-3}}\right) \times \frac{1}{112.48} = 74.09 \ m^2$$

$$A_{min} = \sum(A_i) = 1,153.84 + 197.54 + 2,642.47 + 7,862.81 + 977.37 + 74.09 = 12,908.12 \ m^2$$

Second Case: After Design

In order to estimate the required surface area of heat exchangers, after designing the heat network, we shall design the grid diagram (Figure 6.8) of the process unit which it was already plotted for this unit in previous sections.

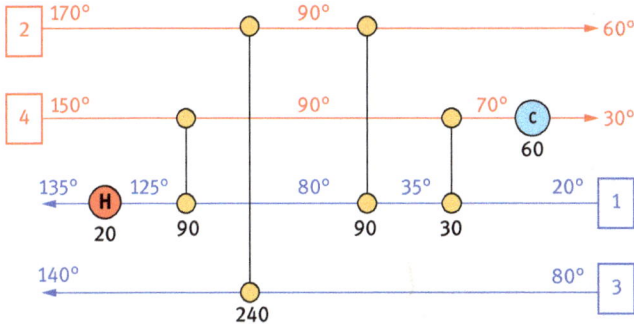

Figure 6.8: Grid diagram for Example 6.2.

By considering the process data in the grid diagram for this unit, we can calculate the surface area of heat exchangers:

$T_{cout} = 125\ °C \qquad T_{cin} = 80\ °C$

90 MW

$T_{hin} = 150\ °C \qquad T_{hout} = 90\ °C$

$$LMTD = \frac{(150 - 125) - (90 - 80)}{\ln\dfrac{150 - 125}{90 - 80}} = 16.37$$

$$A_1 = \left(\frac{1}{5 \times 10^{-3}} + \frac{1}{3.5 \times 10^{-3}}\right) \times \frac{90}{16.37} = 2,670.39\ m^2$$

$T_{cout} = 140\ °C \qquad T_{cin} = 80\ °C$

240 MW

$T_{hin} = 170\ °C \qquad T_{hout} = 90\ °C$

$$LMTD = \frac{(170 - 140) - (90 - 80)}{\ln\dfrac{170 - 140}{90 - 80}} = 18.2$$

$$A_2 = \left(\frac{1}{4 \times 10^{-3}} + \frac{1}{4.5 \times 10^{-3}}\right) \times \frac{240}{18.2} = 6,227.11\ m^2$$

$T_{cout} = 80\ °C \qquad T_{cin} = 35\ °C$

90 MW

$T_{hin} = 90\ °C \qquad T_{hout} = 60\ °C$

$$LMTD = \frac{(60 - 35) - (90 - 80)}{\ln\dfrac{60 - 35}{90 - 80}} = 16.37$$

$$A_3 = \left(\frac{1}{4.5 \times 10^{-3}} + \frac{1}{3.5 \times 10^{-3}}\right) \times \frac{90}{16.37} = 2,792.56 \ m^2$$

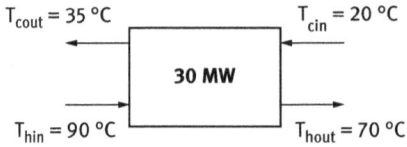

$T_{cout} = 35 \ °C$ $T_{cin} = 20 \ °C$

30 MW

$T_{hin} = 90 \ °C$ $T_{hout} = 70 \ °C$

$$LMTD = \frac{(90 - 35) - (70 - 20)}{\ln\dfrac{90 - 35}{70 - 20}} = 52.46$$

$$A_4 = \left(\frac{1}{3.5 \times 10^{-3}} + \frac{1}{5 \times 10^{-3}}\right) \times \frac{30}{52.46} = 277.76 \ m^2$$

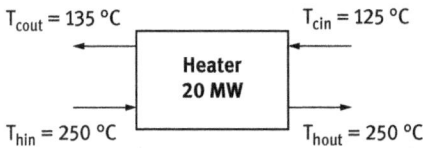

$T_{cout} = 135 \ °C$ $T_{cin} = 125 \ °C$

Heater
20 MW

$T_{hin} = 250 \ °C$ $T_{hout} = 250 \ °C$

$$LMTD = \frac{(250 - 125) - (250 - 135)}{\ln\dfrac{250 - 125}{250 - 135}} = 119.93$$

$$A_{heater} = \left(\frac{1}{3.5 \times 10^{-3}} + \frac{1}{6 \times 10^{-3}}\right) \times \frac{20}{119.93} = 75.44 \ m^2$$

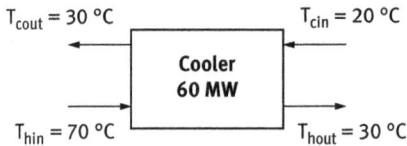

$T_{cout} = 30 \ °C$ $T_{cin} = 20 \ °C$

Cooler
60 MW

$T_{hin} = 70 \ °C$ $T_{hout} = 30 \ °C$

$$LMTD = \frac{(70 - 30) - (30 - 20)}{\ln\dfrac{270 - 30}{230 - 20}} = 21.64$$

$$A_{cooler} = \left(\frac{1}{5 \times 10^{-3}} + \frac{1}{3.75 \times 10^{-3}}\right) \times \frac{60}{21.64} = 1,293.9 \ m^2$$

$$A_T = \sum (A_i) = 2,670.39 + 6,227.11 + 2,792.56 + 277.76 + 75.44 + 1,293.9 = 13,337.16 \ m^2$$

6.4 Estimating the Number of Shells and Heat Transfer Area of Heat Exchanger Network

For industrial applications, if the target temperature of hot stream is higher than the target temperature of cold stream, we would use a simple design consisting of one shell in heat exchanger. The 1–1 shell and tube exchangers have a minimum level of heat transferred. Therefore, the 1–2 shell and tube heat exchanger is mostly used for industrial purposes. Because in this type of exchanger by increasing the surface area and the number of shells, the outlet temperature of cold stream will be greater than the outlet temperature of hot stream. In this regard, the obtaining temperature cross will cause an increase in the amount of energy recovery as well as in developing the energy efficiency. To consider this issue in design process and improve the exchanger's efficiency, we shall use correction factor.

6.4.1 Trial-and-Error Methods and Rule of Thumb for Estimating the Number of Shells in Exchanger

In case that multishell configuration is required for increasing the efficiency of exchanger in the network, designers typically use trial-and-error methods and rule of thumb, such as the Kern method, which can estimate the correction factor for each shell and obtain an appropriate design for the heat exchanger. In this way, we assume to have one shell, and then determine the relevant correction factor [14, 15].

If the value of correction factor is not acceptable for one shell, more quantity of shell will be considered in series, until a proper value of correction factor is gained in every shell (Figure 6.9). In this method, the surface area of all shells and correction factors in converters would be the same and equal, and then the value of $F_T \geq 0.75$ would be approved. Because in the value range of $F_T \geq 0.75$, the maximum temperature cross could be estimated (Figure 6.10).

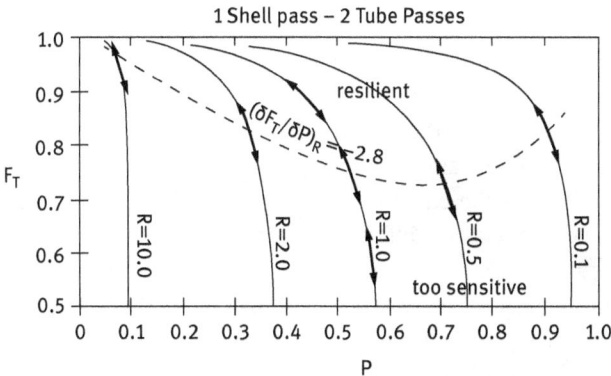

Figure 6.9: Diagram of correction factor variation in comparison to heat efficiency in heat exchanger 2–1.

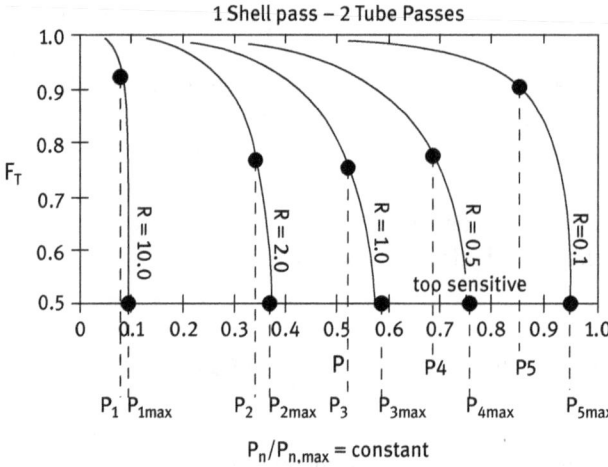

Figure 6.10: $F_T - P$ diagram by calculating vertical asymptotic lines (P_{max}) in heat exchanger 2–1.

An analytical equation for calculating P_{max} (vertical asymptote) was proposed by Mitson in 1984 [16]. In this relation, for every value of R, there will be a maximum value for P (P_{max}), for which in return F_T tends to $-\infty$:

$$P_{max} = \frac{2}{R + 1 + \sqrt{R^2 + 1}}$$

The values greater than P_{max} are not valid; therefore, the practical designs are limited to certain P_{max}:

$$P = X_P \cdot P_{max}, \quad 0 < X_P < 1$$

X_P is a constant defined by the designer and is considered for the purpose of the exchanger design. For instance, with the geometric location for the assumed values of $X_p = 0.9$ in Figure 6.11, this value is plotted on the $P–F_T$ graph. Obviously, the greater amount of F_T and R is acceptable and the lower points are not acceptable.

The exchangers that have been designed according to this method are not sensitive to changes in flow rates and temperature differences. But in order to have reliable designs that can withstand more temperature changes in the inlet of cold stream, the value of angular coefficient or X_P must be decreased. Hence, in general, the reliability of an exchanger depends on the designer's choice of parameters such as X_P.

In case that the correction factor is too low or temperature cross is too high, the 1–2 shell and tube exchanger will be sensitive to temperature changes, and designer would have to increase the number of shells or consider another type for shell to design and make the exchanger properly. Figures 6.12 and 6.13 are turning out the conventional methods of reducing the temperature cross in shells and consequently in heat exchangers.

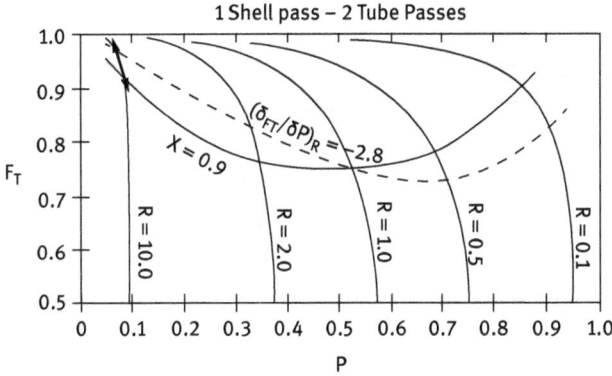

Figure 6.11: $F_T - P$ diagram by calculating $X_p = 0.9$ graph in heat exchanger 2–1.

By using the P_{max} and P relations as well as the following equations, another formula could be achieved to determine the number of shells in an exchanger:

$$N_{shell} = \frac{\left(\frac{1 - RP}{1 - P}\right)}{\ln W}, \quad R \neq 1$$

$$W = \frac{R + 1 + \sqrt{R^2 + 1} - 2RX_P}{R + 1 + \sqrt{R^2 + 1} - 2X_P}$$

$$N_{shell} = \frac{\left(\frac{P}{1-P}\right)\left(1 + \frac{\sqrt{2}}{2} - X_P\right)}{X_P}, \quad R = 1$$

6.4.2 Stepping-Off Method for Estimating the Number of Shells in Heat Exchangers

In this method, we first plot the $(T-H)$ diagram for hot and cold streams, as a designer assumes a specific value for X_P, then the amount of R and P for each shell will be calculated according to the supply and target temperatures of hot and cold streams [17]. Therefore, the values for slope coefficient of lines and the unknown temperature in every shell will be determined:

$$P_{1,2} = \frac{\Delta T_H}{\Delta T_1 + \Delta T_C}, \quad \Delta T_1 = T_{hi} - T_{co}$$

$$R = \frac{\Delta T_C}{\Delta T_H}, \quad \Delta T_H = \frac{\Delta T_1 . P_{1,2}}{1 - R \, P_{1,2}}$$

To determine the number of required shells, we draw a horizontal line from the cold stream target temperature to intersect the hot stream graph in the composite

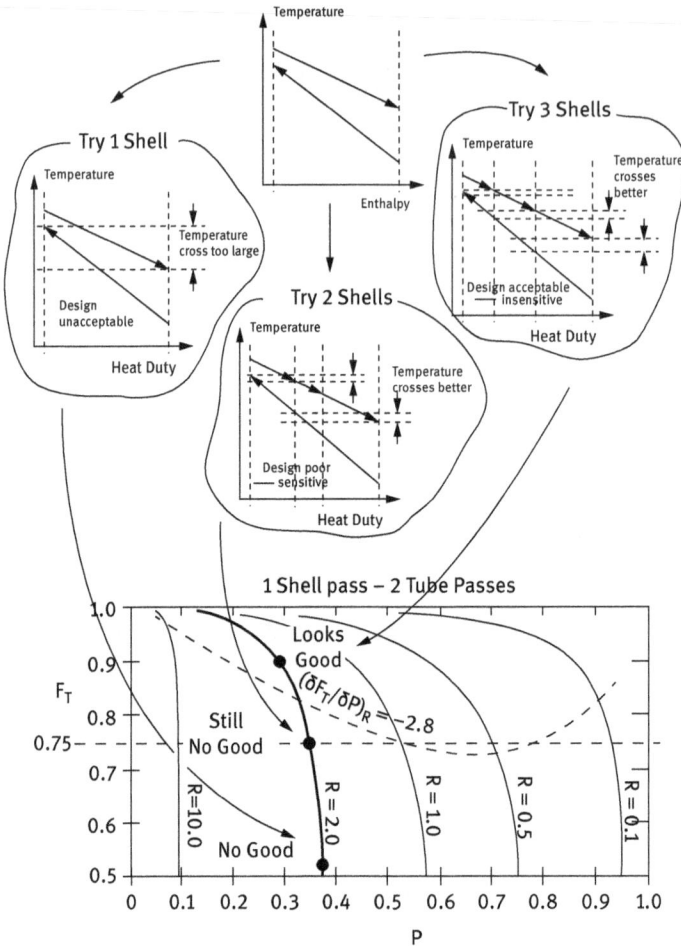

Figure 6.12: The effect of increasing the number of shells on the R curve in $F_T - P$ diagram for 2–1 heat exchanger.

curve, and then draw a vertical line to intersect the cold stream graph again. In this regard, in every situation, a horizontal line is drawn between two hot and cold curves, and step by step, the space between hot and cold flow curves will be completed as stepping-off from the beginning to the end. We keep continuing as long as the horizontal line drawn from the cold stream curve no longer intersects the curve of hot stream and actually it is not possible to plot any other horizontal line between the space of the two curves. Hence, the horizontal lines in the final created graph represent the required number of shells to design a desired heat exchanger, which is often estimated more than the number of shells required by the exchanger to be more reliable.

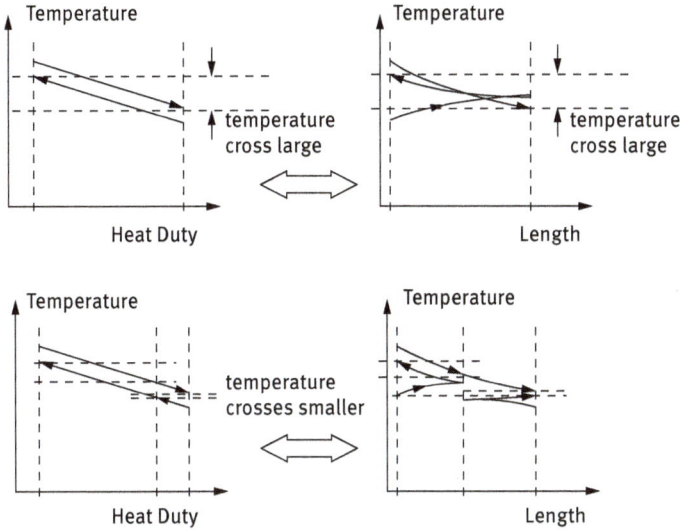

Figure 6.13: The effect of increasing and decreasing the temperature cross on the diagrams related to heat exchangers 2–1.

As an example in the exchanger with a hot stream which is supposed to lose its temperature from 410 to 110 °C, and a cold stream whose temperature must rise from 0 to 360 °C, $R = 1.2$ and $P = 0.73$ will be calculated. By assuming $X_P = 0.9$ and $F_T = 0.75$, the number of shells for this exchanger can be estimated through the stepping-off method. As shown in Figure 6.14, the number of shells is 4.

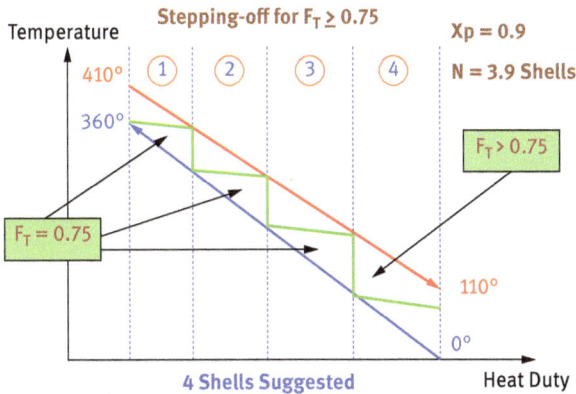

Figure 6.14: Estimating the number of shells in a heat exchanger through the stepping-off method.

There is also a simpler algorithm for determining the minimum number of shells. In this way, same as the surface targeting algorithm, we divide the composite curve into intervals which consist of enthalpies. So the number of shells will be equal to

$$N_{\text{shell}} = \sum_{k}^{\text{Intervals}} N_k(S_K - 1)$$

where N_{shell} is the number of all shells, N_k is the actual number of shells based on the temperature in enthalpy range and S_K is the number of streams in the enthalpy range.

Chapter 7
Estimating the Cost of Heat Exchanger Network

7.1 Energy Cost Targeting for Heat Exchanger Network

Considering the value of $\Delta T_{\min}$ in each network design, in order to estimate the energy cost by using a grid diagram, we calculate $Q_{H_{\min}}$ and $Q_{C_{\min}}$, and then multiply them by the related utility costs, respectively. (Assume that only one type of hot and cold utility has been used.) [18]

We would examine the calculation of capital cost in the following section:

$$\text{Energy costs} = \sum Q_H \times \$_{Q_H} + \sum Q_C \times \$_{Q_C}$$

where Q_H is the heating load of hot utility, Q_C is the cooling load of cold utility, $\$_{Q_H}$ is the cost for selected hot utility and $\$_{Q_C}$ is the cost for selected cold utility.

7.2 Capital Cost Targeting for Heat Exchanger Network

The initial capital costs of the heat exchanger network include the cost for manufacturing, purchasing, installation of the exchangers and supplying the other equipment in order to run the heat exchanger network, which might vary in every project based on the project's objectives. But the cost for purchasing the heat exchangers plays an important role in capital cost targeting [19, 20]. To determine the cost of purchasing an exchanger, we can use the following equation:

$$a + bA^C = (\$)\,\text{Cost for purchasing an exchanger}$$

where a is the coefficient of installation costs, b is the coefficient of exchanger material, c is the coefficient that is a function for the motion of two fluids inside the exchanger and A is the surface area of each exchanger.

Depending on the type of industrial unit and fluids used in process and installation lines, different metals and alloys such as titanium, carbon steel and stainless steel are being used to produce exchangers. Therefore, the cost of manufacturing exchangers will change according to the heat transferring surface area (number of shells) and the material of shells and tubes.

https://doi.org/10.1515/9783110786323-007

Different models have been presented for the function of exchanger cost, depending on the shell and tube material, which we introduce one of them in the following table:

No.	Cost function	Shell and tube material
1	Cost (\$) = 30,800 + 750 $A^{0.81}$	CS/CS
2	Cost (\$) = 30,800 + 1,644 $A^{0.81}$	SS/SS
3	Cost (\$) = 30,800 + 4,407 $A^{0.81}$	TI/TI
4	Cost (\$) = 30,800 + 1,339 $A^{0.81}$	CS/CS
5	Cost (\$) = 30,800 + 3,349 $A^{0.81}$	CS/TI
6	Cost (\$) = 30,800 + 3,749 $A^{0.81}$	SS/TI

In order to determine the cost price of each exchanger, according to the calculation patterns in this table, we shall finally multiply the numbers obtained from the earlier equations by 3.5.

In this regard, the capital cost of the network could be calculated as follows:

$$\text{Capital cost of heat exchanger network} = N_{\text{shells}}\left(a + b\left(\frac{A_{\text{min}}}{N_{\text{shells}}}\right)^{c}\right)$$

In this equation, it is assumed that the surface area of all shells is equal to each other, and the phrase in parentheses actually indicates the cost for one shell.

It is noteworthy that in the mentioned equations, for estimating the minimum surface area of all heat exchangers of network (A_{min}) in the last chapter, it was assumed that all heat exchangers are the same, while most of the time exchanger networks are being selected based on the type of fluids and network capital costs in the process. Therefore, exchangers will not always be used with the same material.

In case that different combinations are used for manufacturing exchangers, the equation of

$$A_k = \frac{1}{(\text{LMTD})_k (F_T)_k}\left(\sum_{i=0}^{n}\frac{Q_i}{h_i}\right)$$

will change as follows:

$$A_k = \frac{1}{(\text{LMTD})_k (F_T)_k}\left(\sum_{i=0}^{n}\frac{Q_i}{\varnothing_i h_i}\right)$$

In this relation, $\varnothing_i$ is a kind of coefficient for the exchanger material and in order to calculate the value for it, we have to choose an exchanger as a reference. (This exchanger has been made with the largest portion in the heat exchanger network for the process unit.) Then through the below equation, we will determine $\emptyset_i$, and we consider the remaining exchangers the same as what we have chosen as a reference:

$$\emptyset = \left[\left(\frac{b_1}{b_2} \right)^{\frac{1}{c_1}} \times A^{1 - \frac{c_2}{c_1}} \right]$$

In this equation that is presented to estimate $\emptyset_i$, the values for b_1 and c_1 are relating to the reference exchanger, and the amount for b_2 and c_2 belong to the dissimilar exchanger with the reference.

7.3 Total Cost Targeting for Heat Exchanger Network

Finally, in order to calculate the total cost of the heat exchanger network, we shall determine the capital costs and energy costs separately, and then sum these values together.

Total cost of heat exchanger network = Energy costs + Capital costs

7.4 Estimating the Minimum Allowable Temperature Difference in Heat Exchangers $(\Delta T_{min})_{opt}$

As you know, ΔT_{min} is a function of capital costs and energy costs, so the optimal value is where the total cost is minimized. To calculate $(\Delta T_{min})_{opt}$, we shall stick to the following instructions:

We assume an initial value for ΔT_{min}.

1. Draw the cascade diagram and with the help of it we determine pinch points $Q_{H_{min}}$ and $Q_{C_{min}}$.
2. Draw the general composite curve and by using it, we choose the most appropriate hot and cold utilities.
3. Calculate the energy cost through the following equation:

$$\text{Energy costs} = \sum Q_H \times \$_{Q_H} + \sum Q_C \times \$_{Q_C}$$

4. Draw the composite curve and by using it, estimate the minimum required surface area.
5. Determine the number of required number of shells in the exchanger network.
6. Calculate the capital costs of heat exchanger network through the following equation:

$$N_{shells}\left(a + b\left(\frac{A_{min}}{N_{shells}}\right)^{c}\right) = \text{Capital costs of heat exchanger network}$$

7. Calculate the total cost of a heat exchanger network.
   Total cost of heat exchanger network = Capital costs + Energy costs.
8. Plot the points regarding energy costs, capital costs and total cost on the diagram of cost – ΔT_{min}, in terms of assumed ΔT_{min}.
9. We return to the first step and specify a new ΔT_{min}, and for the considered ΔT_{min}, we repeat all the above steps. By following these instructions again, the curve of energy costs, capital costs and total cost is plotted in terms of ΔT_{min}. At the end, $(\Delta T_{min})_{opt}$ is where the total cost of heat exchanger network would be estimated as minimum or optimum.

Today, by developing the technology, computer software has been provided to the aid of engineering sciences, which makes long and time-consuming calculations regarding $(\Delta T_{min})_{opt}$, much more simple and accurate [21, 22].

Chapter 8
Furnaces, Heat Engines, Heat Pumps and Distillation Tower

8.1 Furnaces

If we need hot utility with a high temperature, in order to supply $Q_{H_{min}}$, we will use the amount of heat resulting from combustion of fuel in furnaces.

The performance of furnaces is that the fuel and air enter into combustion cylinder and after combustion operation, the fuel gases will leave the cylinder with a temperature (TFT) which is equivalent to the flame temperature. These gases will be used as a hot utility and will provide the heat required by the process and then leave the furnace with the temperature of stack point (T_{stack}) (Figure 8.1).

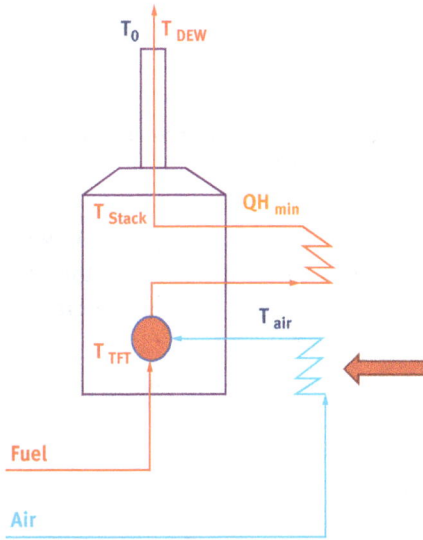

Figure 8.1: Schematic of furnace performance.

The air often gets preheated before entering into the combustion cylinder, causing the flame temperature to rise.

The stack temperature has two limitations (Figure 8.2):

- First, the stack temperature should not be lower than the dew point of sulfur dioxide. Otherwise, due to the presence of sulfur in the fuel, acid corrosion will occur ($T_{stack} \geq T_{DEW}$).
- Second, the stack temperature should not be lower than the pinch temperature. Otherwise, the required heat by a part of process cannot be supplied ($T_{stack} \geq T_{pinch}$).

https://doi.org/10.1515/9783110786323-008

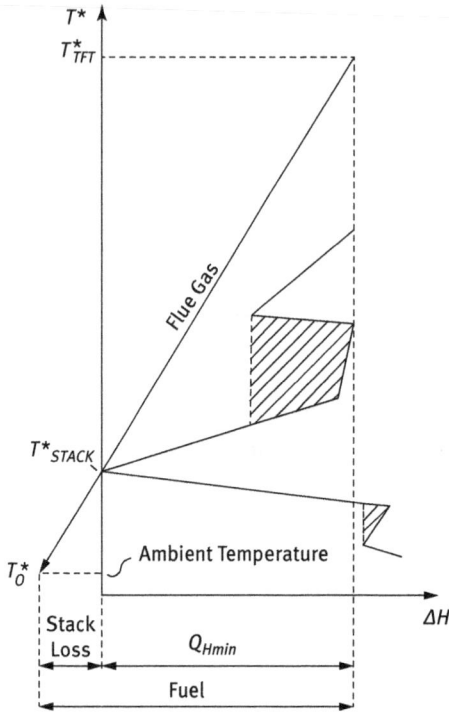

Figure 8.2: General composite curve (GCC) for furnace performance.

Therefore, if the pinch temperature is higher than the dew point temperature (Figure 8.3), the stack temperature will be equal to the pinch temperature, and if the dew point temperature is greater than the pinch temperature, the stack temperature will be equal to the dew point temperature.

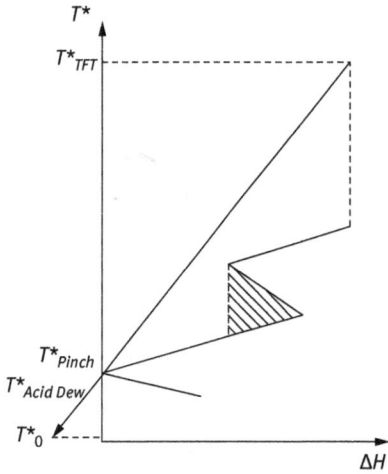

Figure 8.3: General composite curve (GCC) of furnace if $T_{pinch} > T_{DEW}$.

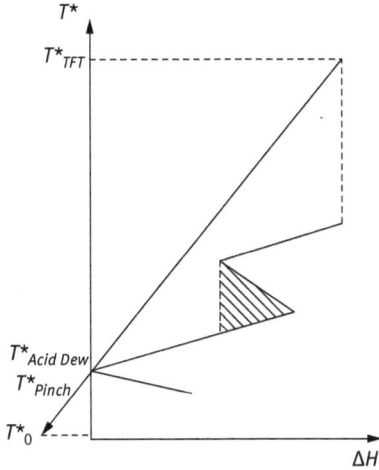

Figure 8.4: General composite curve (GCC) for furnace if $T_{pinch} < T_{DEW}$.

There are three significant points to design furnaces:

1. By increasing the fuel and combustion, the amount of energy can rise and vice versa.
2. By increasing the temperature of preheated air, the flame temperature will increase and the heat loss rate will decrease.
3. The furnaces usually work with 5–20% of excess air, which depends on designing of furnaces and flares. Increasing the excess air will make better combustion but on the other side will decrease the temperature of flame. Therefore, the rate of excess air has an optimum value that by considering this, the furnace could have better and optimum performance.

8.2 Heat Engine (H.E.)

So far, we have examined the optimal use of thermal energy in the process, now at this section we would explain about the heating power.

Thermal power is actually another form of energy that is much more valuable than thermal energy.

The thermodynamic model of heat engines is simply shown in Figure 8.5.

The performance of heat engine is to generate work by receiving energy from a heating source and excrete some energy to the cooling source at a lower temperature. Therefore, the equations relating to the function of heat engines are as follows:

$$w = Q_1 - Q_2$$

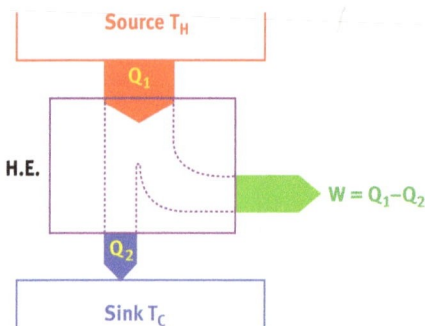

Figure 8.5: Thermodynamic model of heat engine performance.

$$W_{Max} = Q_1 \left(1 - \frac{T_C}{T_H} \right)$$

$$w_{Real} = W_{max} \times \eta_m$$

where W is the work done by the heat engine, W_{Max} is the maximum work that a heat engine can do, W_{Real} is the real work done by the heat engine, Q_1 is the amount of energy that a heat engine receives from the heating source, Q_2 is the amount of energy that a heat engine dissipates through the cooling source, T_H is the heating source temperature, T_C is the cooling source temperature and η_m is the heat engine efficiency.

Heat engines will be divided into two categories: steam turbines and gas turbines.

In steam turbine, the turbine blades are driven by steam at a high-pressure state, while in a gas turbine this action is done via combustion gases.

Now we need to recognize, where is the right place to locate the heat engine in a process?

To answer this question, we examine three possible modes: above the pinch, below the pinch and across the pinch.

8.2.1 Placement of Heat Engine Across the Pinch Point

As we know, the above area of pinch is the receiver of energy, and heat engine needs this energy to generate work. Therefore, the heating load of the heater will increase based on the required energy by turbines. Moreover, we know that in the lower region of pinch, some energy is dissipated through a cold utility, and heat engine delivers some energy to the cold utility as well. Hence, the cooling load of the cooler will increase (Figure 8.6).

In this case, no integration has occurred. Because in order to generate work as W, some energy is received from hot utility and some amount would be dissipated through the cold utility. So it will not be a suitable place to design and place a turbine across the pinch point.

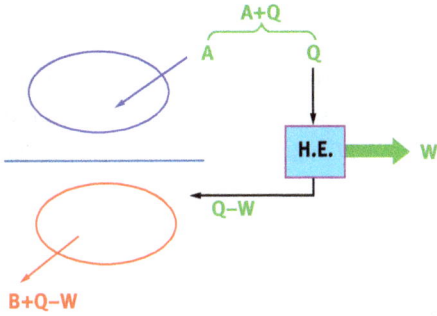

Figure 8.6: Thermodynamic model of heat engine placement across the pinch point.

8.2.2 Placement of Heat Engine Above the Pinch

As it was explained, the above region of pinch point is actually the receiver of energy. If we assume that the received amount of energy by the network at this area is equal to A, the heat engine requires energy equal to Q in order to produce work as much as W. On the other hand, according to Figure 8.7, the heat engine would bring some energy as $Q-w$ back to the process. In this case, the heating load of the heater increases by W, but the cooling load of the cooler will not change. Hence, this process eventually leads to generate work as W.

Therefore, in a simpler term, we consume as much energy as Q, but the same amount of W will be produced in the network. So 100% of energy conversion to work has occurred. This mode is practical and very ideal.

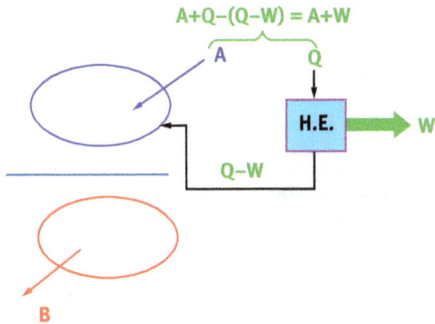

Figure 8.7: Thermodynamic model of heat engine placement above the pinch.

8.2.3 Placement of Heat Engine Below the Pinch

As mentioned earlier, in the below area of pinch, the system will lose some energy through the cooler. According to Figure 8.8, by designing and placing the heat engine in this region, further to supply the required energy for heat engine, the cooling load of the cooler will be reduced as much as w, and the same amount of work will be generated. This intends that without making any changes in the heating load of the heater, some work is produced and also the cooling load of the cooler has been reduced, which is theoretically very ideal. However, the noticeable point is that there would never be a steam stream which is capable of moving the turbine blades in the lower area of pinch. Hence, the lower region of pinch is not a good place to design and locate the heat engine.

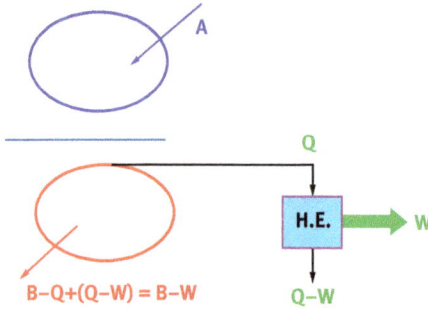

Figure 8.8: Thermodynamic model of heat engine placement in the lower area of pinch.

According to the explanations, it can be concluded that the best place to design and locate turbine is at the upper region of the pinch (Figures 8.9 and 8.10).

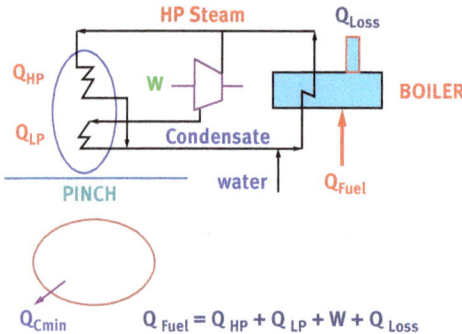

Figure 8.9: Schematic design of heat engine in the upper region of pinch.

8.3 Heat Pump (H.P.)

The performance of heat pumps is opposite to that of heat engines and turbines, as the heat pumps that receive some work will transfer the energy from the cold source to the hot source. Hence, the equations relating to the performance of heat pumps are as follows:

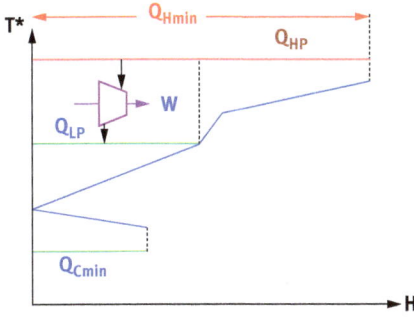

Figure 8.10: General composite curve regarding the heat engine placement in the upper region of pinch.

$$Q_1 = Q_2 + w$$

$$\text{COP} = \frac{Q_2}{w}$$

$$\text{COP}_{h.p} = \frac{T_H}{T_H - T_C}, \quad \text{COP}_{ref} = \frac{T_C}{T_H - T_C}$$

$$W_{hp} = \frac{Q_H}{\eta_m (\text{COP})_{h.p}}, \quad W_{hp} = \frac{Q_C}{\eta_m (\text{COP})_{ref}}$$

where Q_1 is the amount of energy which heat pumps will transfer to the hot source, Q_2 is the amount of energy which heat pumps will receive from the cold source, T_H is the hot source temperature, T_C is the cold source temperature, η_m is the heat pump efficiency and COP is the coefficient of performance.

According to these equations, it can be concluded that as the temperature difference between hot and cold utilities decreases, then the heat pump performance will be better. The thermodynamic model of heat pumps is shown simply in Figure 8.11.

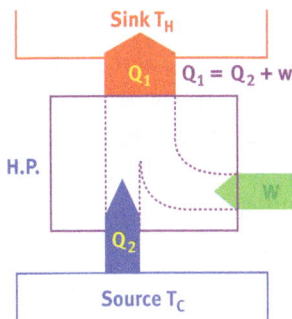

Figure 8.11: Thermodynamic model of heat pump performance.

Now we examine the appropriate location to place the heat pump in a heat network design (upper area of pinch, across the pinch and pinch point).

8.3.1 Placement of Heat Pump Across the Pinch Point

As we know that the lower area of pinch will dissipate some energy through the cooler, the heat pump will receive this energy; hence, the cooling load of cooler will increase as much as Q.

The upper region of pinch is the receiver of energy, and the heat pump will transfer some energy to the heat source. Therefore, the heating load of the heater will decrease (Figure 8.12).

In other words, by using some work as much as W, the heating and energy load are reduced by $Q + W$ from the heater and Q from the cooler. Therefore, across the pinch would be a suitable choice to design and place the heat pumps.

The significant point is that if the temperature difference between two sources gets very large, the required value of work will be so high; hence, it would not be possible to use the heat pump in the process and heat exchanger's network.

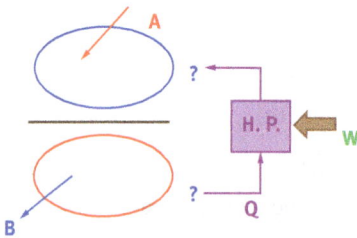

Figure 8.12: Thermodynamic model of heat pump placement across the pinch point.

8.3.2 Placement of Heat Pump Above the Pinch

The upper region of pinch is the receiver of energy, if we assume that the amount of energy received by the network at this area is equal to A, and the heat pump will transfer some energy as much as $Q + W$ to the hot source; on the other hand, it will receive some energy value as Q from the hot source. Therefore, at the end, the heating load of the heater will be reduced as much as W. In other words, some works have been converted into the heat energy which is not economically efficient, since the work is more valuable than the heat energy. Hence, the lower area of pinch would not be a good choice to design and place the heat pump.

8.3.3 Placement of Heat Pump Below the Pinch

As explained earlier, some energy will be dissipated through the cold source. If we assume that this amount is equal to Q, the heat pump at this region will transfer some energy as much as $W + Q$ to the cold source, and it will receive some energy as Q from the cold source (Figure 8.13). Finally, the value of W will be added to the

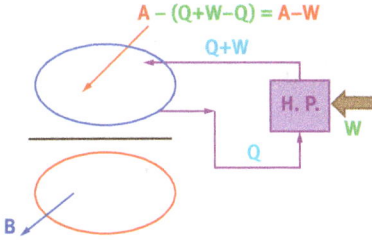

$A - (Q+W-Q) = A-W$
$Q+W$
H. P.
W
Q
B

Figure 8.13: Thermodynamic model of heat pump placement above the pinch.

energy load of the cooler. On the other hand, the heat pump has received some work as W, which means that the value of W has been consumed and the same exact amount is added to the energy load of the cooler (Figure 8.14). Hence, the lower area of pinch is not appropriate to place and design the heat pump.

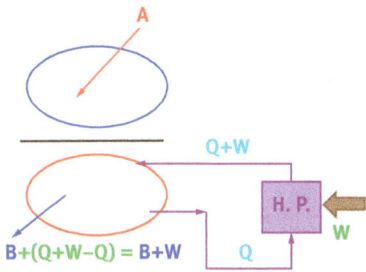

A
$Q+W$
H. P.
W
$B+(Q+W-Q) = B+W$
Q

Figure 8.14: Thermodynamic model of heat pump placement below the pinch.

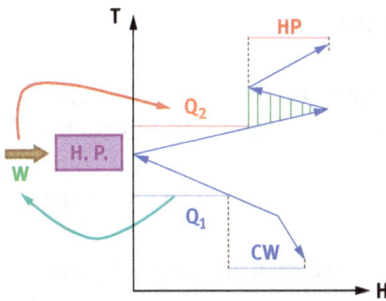

T
HP
Q_2
H. P.
W
Q_1
CW
H

Figure 8.15: General composite curve (GCC) for the heat pump placement across the pinch point.

So according to the performed studies, we conclude that the best place to allocate and design the heat pumps is across the pinch point (Figure 8.15).

8.4 Distillation Tower

As we know that the reboiler receives heat energy in a distillation tower, the condenser delivers some heat and the distillation tower performs separation. So we can say that the distillation tower is a type of heat engine, which means that the distillation tower and heat engine are thermodynamically considered the same.

The most heating and cooling load is consumed by the condenser and reboiler, and both of them will operate at constant temperatures (Figure 8.16).

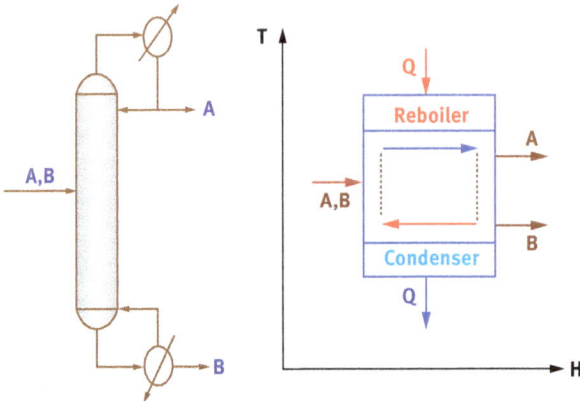

Figure 8.16: Overview of the distillation tower.

Now we examine the appropriate place of heat pump in the heat network design (above the pinch, below the pinch and across the pinch).

8.4.1 Placement of Distillation Tower Across the Pinch Point

As explained earlier, the upper area of pinch is the receiver of energy. The reboiler requires energy as Q_R; therefore, the heating load of the heater will increase by Q_R. On the other hand, the cooling load of the cooler will increase as much as Q_C in the below region of pinch. Hence, at this state no energy will be saved and it just restricts our freedom to design the proper network.

So across the pinch will not be a suitable place to design and locate the distillation tower.

8.4.2 Placement of Distillation Tower Above the Pinch

At the above region of pinch, distillation tower receives the heat energy equal to Q_R, and it brings back the heat as much as Q_C to the process. If we assume that the required amount of energy by the heater is A (Figure 8.17), after locating the distillation tower in the upper area of pinch, this value will be equal to $A + Q_R - Q_C$ and the cooling load of the cooler will not change.

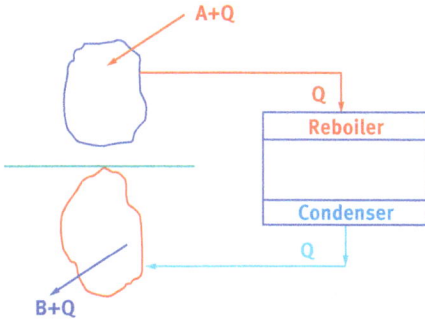

Figure 8.17: Thermodynamic model of distillation tower placement across the pinch point.

Q_R and Q_C are often close to each other. If the values of them are same, then the heating load of the heater will also remain the same.

According to the above explanations, the upper area of pinch could be a suitable place in order to design and locate the distillation tower (Figure 8.18).

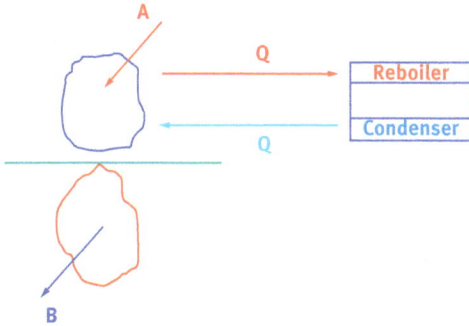

Figure 8.18: Thermodynamic model of distillation tower placement above the pinch.

8.4.3 Placement of Distillation Tower Below the Pinch

Distillation tower in the lower area (Figure 8.19), which is same as the upper area of pinch, will receive the heat energy equal to Q_R and it brings back the value of Q_C into the process. If we assume that B is the amount of energy in which the process initially brings to the cooler, this would be equal to $B - Q_R + Q_C$, and the heating load of the heater will remain the same.

Q_R and Q_C are often close to each other and equal, so the cooling load of the cooler will also remain the same.

Therefore, the lower area of pinch could be a suitable place in order to design and locate the distillation tower.

Hence, as a conclusion and according to the explanations, the upper or lower regions of pinch would be suitable parts to design and place the distillation towers.

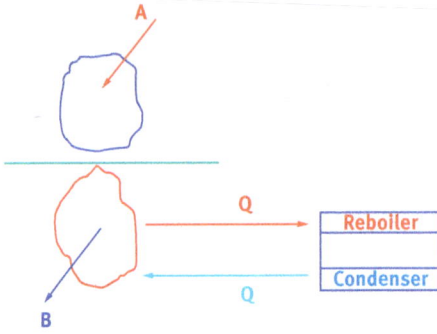

Figure 8.19: Thermodynamic model of distillation tower placement below the pinch.

References

[1]   Bowman, R.A. 1936. Mean temperature difference correction in multipass exchangers. *Industrial and Engineering Chemistry*, 28(5), 541–544.

[2]   Bowman, R.A., Mueller, A.C. and Nagle, W.M. 1940. Mean temperature difference in design. *Transactions ASME*, 62(4), 283–294.

[3]   Binosi, D. and Papavassiliou, J. 2009. Pinch technique: Theory and applications. *Physics Reports*, 479(1–6), 1–152.

[4]   Ciric, A.R. and Floudas, C.A. 1990. A mixed integer nonlinear programming model for retrofitting heat-exchanger networks. *Industrial & Engineering Chemistry Research*, 29(2), 239–251.

[5]   Douglas, J.M. 1988. Conceptual Design of Chemical Processes. 1110, New York: McGraw-Hill.

[6]   Gundersen, T., 2000. A process integration primer – Implementing agreement on process integration. *Trondheim, Norway: International Energy Agency, SINTEF Energy Research*, 34–47.

[7]   Hohmann, E., 1984. Heat-Exchange Technology, Network Synthesis. *Kirk-Othmer Encyclopedia of Chemical Technology*.

[8]   Hohmann, E.C., 1971. *Optimum networks for heat exchange* (Doctoral dissertation, University of Southern California).

[9]   Kemp, I.C. 2011. Pinch analysis and process integration: A user guide on process integration for the efficient use of energy. Elsevier.

[10]   Kern, D.Q. 1950. Process Heat Transfer. McGraW-Hill, New York.

[11]   Kesler, M.G. and Parker, R.O. 1969. Optimal networks of heat exchange. *Chemical Engineering Progress Symposium Series*, 65(92), 111–120.

[12]   Linnhoff, B. and Hindmarsh, E. 1983. The pinch design method for heat exchanger networks. *Chemical Engineering Science*, 38(5), 745–763.

[13]   Linnhoff, B. and Flower, J.R. 1978. Synthesis of heat exchanger networks: I. Systematic generation of energy optimal networks. *AIChE Journal*, 24(4), 633–642.

[14]   Linnhoff, B., 1998. Introduction to Pinch Technology. Linnhoff March Limited, Targeting House Gadbrook Park Northwich, Cheshire, UK.

[15]   Linnhoff, B., Mason, D.R. and Wardle, I. 1979. Understanding heat exchanger networks. *Computers & Chemical Engineering*, 3(1–4), 295–302.

[16]   Mitson, R.J., 1984. Number of shells vs number of units in heat exchanger network design (Doctoral dissertation, Univ. of Manchester, Inst. of Sci. and Technol).

[17]   Pho, T.K. and Lapidus, L. 1973. Topics in computer-aided design: Part II. Synthesis of optimal heat exchanger networks by tree searching algorithms. *AIChE Journal*, 19(6), 1182–1189.

[18]   Smith, R. 2005. Chemical Process: Design and Integration. John Wiley & Sons.

[19]   Taborek, J. 1979. Evolution of heat exchanger design techniques. *Heat Transfer Engineering*, 1(1), 15–29.

[20]   Tjoe, T.N. and Linnhoff, B. 1986. Using pinch technology for process retrofit. *Chemical Engineering*, 93(8), 47–60.

[21]   Umeda, T.A., Itoh, J. and Shiroko, K. 1978. Heat-exchange system synthesis. *Chemical Engineering Progress*, 74(7), 70–76.

[22]   Yee, T.F. and Grossmann, I.E. 1991. A screening and optimization approach for the retrofit of heat-exchanger networks. *Industrial & Engineering Chemistry Research*, 30(1), 146–162.

https://doi.org/10.1515/9783110786323-009

Index

https://doi.org/10.1515/9783110786323-010

www.ingramcontent.com/pod-product-compliance
Lightning Source LLC
Chambersburg PA
CBHW081551220326
41598CB00036B/6639

* 9 7 8 3 1 1 0 7 8 6 3 1 6 *